RECENT ADVANCES IN HAMILTONIAN SYSTEMS

Proceedings of the International Conference on

RECENT ADVANCES IN HAMILTONIAN SYSTEMS

University of L'Aquila,
10–13 June 1986

Edited by

G F Dell'Antonio & B D' Onofrio

World Scientific

Published by

World Scientific Publishing Co Pte Ltd.
P.O. Box 128, Farrer Road, Singapore 9128

Library of Congress Cataloging-in-Publication data is available.

RECENT ADVANCES IN HAMILTONIAN SYSTEMS

ISBN 9971-50-246-1

Printed in Singapore by Kyodo-Shing Loong Printing Industries Pte Ltd.

Preface

In the past few years, much progress has been made in the study of Hamiltonian systems and their periodic trajectories. The purpose of this Conference was to provide a meeting place for scientists who have contributed to these developments, with the aim of creating a stimulus for exchange of information and of bringing techniques and results to the attention of a wider audience.

The number of partecipants stabilized at about forty, a very attentive and stimulating audience.

A substantial part of the Conference centered on the use of topological and variational methods to establish lower bounds to the number of periodic solutions with prescibed energy or with prescribed minimal period.

Other topics treated were bifurcation for dynamical systems, especially in case of resonance, and the rôle of exponential time scales in mechanical systems, a subject related to Arnold's diffusion.

There were eleven main lectures and several shorter contributions. We outline briefly the content and scope of those contributions which appear in these proceedings, grouping them loosely according to the problems analyzed, even when the techniques used are rather different.

Two of the contributions, namely those of V. Benci and F. Pacella are devoted to the description of general techniques which play a rôle in this field of research.

In his contribution, V. Benci gives a rather detailed and self-contained presentation of a new index theory, which extends Conley's theory to non locally compact metric spaces. Definitions and results are presented in detail and most of the proofs are either given or outlined. This index theory has already found applications also in the field of research treated in this Conference (see, e.g. the contribution of D. Fortunato).

The contribution of F. Pacella to this proceedings is also reasonably self-contained. It describes the different procedures which have been developed to derive Morse-type inequalities when the system is invariant under the action of a topological group, in particular when the action is not free. Simple examples are provided to point out the difference in the type of results obtained.

The contribution of C. Chaperon sets the problem of periodic orbits in the context of the intersection of Lagrangean submanifolds.

More specifically, Chaperon studies the number of intersections of a Lagrange immersion j: $L \to T^*M$ with the zero section $\mathcal{O}_M \equiv \{M\} \times \{0\}$. He considers in particular the immersions generated by quadratic phase functions and illustrates some recent work of Sikorav in which the general case is reduced to the special one of diffeomorphisms of T^*M which equal the identity outside a compact set and satisfy some further properties.

For this class of diffeomorphisms, Chaperon outlines the proof of the following result: there are at least N+1 points in $\mathcal{O}_M \cap j(\mathcal{O}_M)$ and at least 2^n if all intersections are transversal (n is the dimension of M).

The contribution by H. Hofer gives a rather extended outline of work done in collaboration with I. Ekeland; a complete account will be published elsewhere. The aim is to estimate the number of periodic trajectories on a prescribed energy surface $S \subset R^{2n}$, of co-dimension one. S is assumed compact, and bounding a convex region containing the origin; it is also assumed that S has non-vanishing Gauss curvature. If Ω is the standard symplectic form in R^{2n}, and ω its restriction to S, then (S,ω) is a contact manifold and one can define "hamiltonian" trajectories on S.

Let S be the set of periodic trajectories.

Ekeland and Hofer prove that one can associate to S a global invariant (related to the homology of the level sets of a function defined on the maps from S^1 to R^{2n} through the Fenchel conjugate of the Hamiltonian H_s) and to every $G \in [S]$ an invariant (related to the phase of the determinant of the tangent map of the hamiltonian flow) and that these invariants cannot be independent, if [S] is finite.

The authors draw then several conclusions on the number of periodic solutions, both in general and under further conditions on S.

In his contribution, A. Ambrosetti establishes the existence of periodic solutions with pre-assigned period T for a mechanical system defined in an open set $\Omega \subset R^N$, $\Omega \neq R^N$, with a potential V which is singular at $\partial\Omega$. When $R^N - \Omega$ is compact and $U \geq 0$ in Ω, under a "strong force" conditions at $\partial\Omega$, and a further technical assumption on the behaviour of U(x) when $|x| \to \infty$, Ambrosetti proves that for any $T > 0$ there are infinitely many periodic solutions with period T.

The contributions of M. Bertotti, V. Coti-Zelati and A. Salvatore treat several aspects of the problem of proving existence of one, two or more T-periodic solutions of a Hamiltonian system when the Hamiltonian is T-periodic, or of a Lagrangean system with a T-periodic potential and forcing term.

The contribution of M. Bertotti establishes the existence of at least two (in some cases, at least three) T-periodic solutions for a mechanical system with a T-periodic potential which is asymptotically linear at the origin and at infinity. Her results, obtained through Morse's inequalities, can be viewed as an extension to this case of the classical fixed-point theorem of Poincaré-Birkhoff.

V. Coti-Zelati studies mechanical systems with a T-periodic potential V(x), x R , under rather general assumptions on the behaviour of V at the origin and at infinity. The results presented are consequences of a general theorem which the author derives using classical

Morse theory and a result of Marino-Prodi on degenerate critical points. Existence of at least one T-periodic solution (or, under further conditions on V, of at least two) is established for potentials of the form $V(x) = \frac{1}{2} x^2 + U(x,t)$, where $|x|^{-1} U_x(x,t) \to 0$ when $|x| \to \infty$, uniformly in t.

In her contribution, A. Salvatore proves existence of at least two T-periodic solutions, which do not differ by time-translation by $K\tau$, $K \in Z$, for a forced mechanical system in which the Lagrangean is τ-periodic, the forcing term is T-periodic and the potential is bounded. Existence is also established for the case in which the forcing term is absent but the potential is singular at the origin.

The contributions of D. Fortunato (reporting on a joint work with V. Benci) and of G. Tarantello treat the problem of subharmonics of prescribed minimal period KT, $K \in Z$, for Hamiltonian system which is T-periodic.

G. Tarantello treats the case of subquadratic Hamiltonians; she establishes a lower bound on the number of such periodic solutions using the dual variational principle, a version of the Borsuk-Uhlam lemma and equivariant Morse theory in the space of 2KT-periodic functions with value in R^{2n} (n is the number of degrees of freedom).

V. Benci and D. Fortunato use the direct variational principle and Benci's index theory (see the contribution of V. Benci to these Proceedings). They treat the case of a Lagrangean of the form $L = L_o - V$ where L_o is the Lagrangean of a system of harmonic oscillators with frequencies $\omega_i \geq 0$, $i = 1 \ldots n$, and V is superquadratic, and give a lower bound to the number of subharmonics of minimal period KT, $K \in Z$.

Both G. Tarantello and V. Benci - D. Fortunato make an assumption which guarantees that H is "truly periodic". Some results are also presented for the case in which this assumption is dropped, and in particular for the autonomous case.

M. Matzeu (reporting on joint work with M. Girardi) treats the autonomous case, under the assumption that the Hamiltonian H can be written as $H = H_2 \neq H'$, where H' is strictly convex and H_2 is quadratic and strictly positive with eigenvalues $\omega_1 \geq \ldots \geq \omega_N > 0$. These authors use a variant of the dual variational principle, taking $J\frac{dz}{dt} - \nabla H_2(Z)$ as linear operator inducing duality. They also exploit a result by Hofer on the value of the Morse index for a critical point of Mountain Pass type. They prove that, for any $T < \frac{2\pi}{\omega_N}$, there exists a periodic solution with minimal period T, and this solution lies outside a ball of radius r(T), where $\lim_{T \to 0} r(T) = +\infty$. Results are also given on the existence of periodic solutions with minimal period in a left neighbourhood of $\frac{2\pi}{\omega_i}$, $i = 1 \ldots N-1$.

The work presented by F. Giannoni gives lower bounds, under different assumptions on the geometrical shape of Ω, on the number of periodic trajectories with prescribed minimal period T for the motion of a material point which moves in a potential field and undergoes elastic collisions at the boundary of a convex billiard $\Omega \subset R^n$.

The contributions of P. Negrini and L. Salvadori refer to bifurcations of dynamical systems when two resonating eigenvalues cross the imaginary axis.

P. Negrini illustrates a way in which the perturbation analysis developed by Kolmogorow - Arnold - Moser for Hamiltonian systems can be used to study the bifurcation of invariant two-dimensional tori into invariant three-dimensional ones for some non-hamiltonian dynamical systems, depending on two parameters. Negrini determines explicitely the curves in parameter space along which bifurcation takes place.

L. Salvadori studies the bifurcation and related stability problems of a one-parameter family of differential systems $R_n(\mu)$ in $\mathbb{R}^n$, in particular for n = 2. It is assumed that $R_2(0)$ is autonomous and that the asymptotic stability of the origin of $R_2(0)$ is preserved under perturbations of order greater than some integer h. A detailed description is given of the different possibilities which can arise depending on the value of h and on resonance and transversality conditions. Included in this contribution is also a study of the existence and stability of periodic solutions.

The contribution of H. Berestycki (reporting on joint work with J.M. Lasry) deals with the existence of periodic orbits for conservative systems. They prove for example that if the mapping $\phi \in C(S^{2n-1}, R^{2n})$ is such that $x \cdot \phi(x) = 0$ and $|\phi(x) - Jx| < \frac{1}{3} x$, then there is at least one periodic orbit of $\dot{x} = \phi(x)$ on S^{2n-1}.

The method rests on: 1) the study of complex eigenvalues for some non-linear problems; 2) a topological formulation via equivariant homotopy theory; 3) a-priori estimates involving a "principal coefficient" method.

Finally, the contribution of G. Benettin belongs to the general area of study of the behaviour on an exponentially long time scale of Hamiltonian systems which are nearly integrable.

Benettin uses the techniques and some of the results of Nekhoroshev to prove the following result: if in a mechanical system with time-independent holonomic bilateral constraints one approximates the constraints by a suitable very strong conservative force, the trajectories of the original system can be well approximated on an exponentially long time scale.

The conference was supported by the CNR through the Committee for Mathematics and by the Ministry of Public Education.

The organizing Committee
(C. D'Antoni, S. De Gregorio,
G. Dell'Antonio, B. D'Onofrio

CONTENTS

RECENT ADVANCES IN HAMILTONIAN SYSTEMS

A NEW APPROACH TO THE MORSE-CONLEY THEORY

Vieri Benci
Istituto di Matematiche Applicate "U. Dini"
Università di Pisa
56100 Pisa
ITALY

0. INTRODUCTION

In this note we present a new approach to the Morse theory which is based on a generalization of the Conley index to non locally compact spaces. The variant of the Morse theory which we obtain seems suitable for applications to nonlinear functional analysis. We refer to a paper in preparation [B] for some of such applications.

Since we are not well acquainted with the very extensive literature on Morse theory, we did not attempt to provide a listing even of the main papers on this subject. We apologize for this to the readers and to the authors.

1. THE HOMOTOPIC INDEX

Let M be a metric space on which a semiflow η is defined i.e. a (continuous) map

$$\eta : \mathbb{R}^+ \times M \to M$$

To appear in the proceedings of the meeting on Hamiltonian Systems held at L'Aquila in June 1986 such that $\eta(0,x) = x$ and $\eta(t_1,\eta(t_2,x)) = \eta(t_1+t_2,x)$; $(t\ ,t\ \in \mathbb{R}^+,\ x \in M)$.

When no ambiguity is possible we will write $x \cdot t$ instead of $\eta(t,x)$.

A semiflow which is defined for every $t \in \mathbb{R}$ is called a flow. If X is any subset of M and T a positive constant we set

$$G^T(X) = G^T(X,\eta) = \{x \in M \mid x\cdot[0,T] \in M\} \cap \{\bigcup_{t \geqq 0} \eta(t,X)\}$$

If η is a flow, clearly we have

$$G^T(X) = \{x \in M \mid x\cdot[-T,T] \in X\} = \bigcap_{t \in [-T,T]} \eta(t,X).$$

Also we set

$$\Sigma = \Sigma(\eta) = [X \subset M \mid X \text{ is closed and } \exists\ T > 0 \text{ s.t. } G^T(X,\eta) \subset \mathring{X}\}$$

where $\mathring{X}$ denotes the interior of X.

<u>Def. 1.1</u> A pair of closed subset of X, (N,N_o) with $N_o \subset N$ is called <u>index pair</u> if

(i) $\overline{N - N_o} \subseteq \Sigma$ ($\bar{A}$ denotes the closure of A)

(ii) N_o is positively invariant with respect to N (i.e. $x \in N_o$ and $x\cdot[0,t] \in N \quad x\cdot[0,t] \in N_o$)

(iii) N_o is an <u>exit</u> set for N (i.e. $x \in N$ and $x\cdot[0,t] \not\subset N \ \ \exists\, t^* \in [0,t]$ such that $x\cdot t^* \in N_o$)

We say that (N,N_o) is an index pair for $X \quad \Sigma$ if

(iv) $\overline{N - N_o} \subset X$ and there exists $T > 0$ such that $G^T(X) \subset \overline{N - N_o}$.

Now it is necessary to recall some concepts from the homotopy theory.

If X is a topological space and A is a closed subset then X/A denotes the spaces obtained by X identifying all the points of A.

Two spaces X/A and Y/B are called homotopic equivalent if there are maps $\phi : X/A \to Y/B$ and $\psi : Y/B \to X/A$ such that $\phi([A]) = [B]$; $\psi([B]) = [A]$ and such that $\phi \circ \psi$ and $\psi \circ \phi$ are homotopic to the identity by homotopies which leave the points [A] and [B] fixed respectively.

The class of all spaces homotopically equivalent to X/A is called homotopy type of X/A and denoted by [X/A].

The homotopy type of X/X is denoted by $\underline{0}$; if X is a contractible space, the homotopy type of X/Φ is denoted by $\underline{1}$. Morever, by convention, we set $\Phi/\Phi = \underline{0}$.

<u>Def. 1.2</u> For $X \in \Sigma$, the homotopy index of X is the homotopy type of an index pair $(N,N_\circ)$ relative to X; in formula we write

$$h(X) = h(X,\eta) = [N/N_\circ]$$

The definition 1.2 makes sense if we prove that

(a) $\forall\ X \in \Sigma$ there exists an index pair $(N,N_\circ)$ for X

(1-1)

(b) if $(N,N_\circ)$ and $(\tilde{N},\tilde{N}_\circ)$ are two index pairs relative to X, then $[N/N_\circ] = [\tilde{N}/\tilde{N}_\circ]$

In order to prove (1-1) some work is necessary. First, we need another notation; for $T > 0$ we set

(1-2) $\quad \Gamma^T(X) = \Gamma^T(X,\eta) = \{x \in G^T(X,\eta) \mid x\cdot[0,T] \cap \partial X \neq \Phi\}$

We need now a technical lemma:

Lemma 1.3 Suppose that $X, Y \in \Sigma$; then

(i) $\quad X \subset Y \to G^T(X) \subset G^T(Y)$ for every $T > 0$

(ii) $\quad T_1 > T > 0 \to G^{T_1}(X) \subset G^T(X)$

(iii) $\quad G^{T_1+T_2}(X) = G^{T_1}(G^{T_2}(X))$

(iv) $\quad$ if $G^T(X) \subset \overset{\circ}{X}$ then $G^{2T}(X) \subset \operatorname{int}[G^T(X)]$

(v) $\quad G^T(X)$ is closed

(vi) $\quad$ if $X \in \Sigma$ then $G^T(X)$ and $\eta(t,X) \in \Sigma$

(vii) $\quad \Gamma^T(X)$ is closed

(viii) $\quad \Gamma^T(X) \subset \partial G^T(X)$

Proof. (i), (ii) and (iii) follow easily from the definition of $G^T(X)$.
(iv) First of all observe that if

(1-3) $\quad x \in G^T(X)$,

then $x \cdot t$ is defined for $t \in [-T,T]$ i.e. we can go back in time up to the point $-T$; and this by the definition of $G^T(X)$.

In order to prove (iv) we argue indirectly and we suppose that there exists $y \in G^{2T}(X) \cap \partial G^T(X)$. Then there exists a sequence $y_n \to y$ such that $Y_n \cdot [-T,T] \not\subset X$. This implies that there exist times $t_n \in [-T,T]$ such that $y_n \cdot t_n \notin X$; we can extract a sequence t_n' such that $t_n' \to \bar{t}$, so we have that $y_n \cdot t_n' \to y \cdot \bar{t} \notin \partial X$. Since $y \in G^{2T}(X)$, $y\cdot[-2T,2T] \subset X$ and so $y \cdot \bar{t} \in G^T(X)$ (since $|\bar{t}| \leqq T$).

And this contradicts our assumption that $G^T(X) \cap \partial X = \Phi$.

(v) We have $G^T(X) = \{\bigcap_{t\in[0,T]} \eta(t,X)\} \cap \{x \in X \mid x\cdot[0,T] \subset X\}$

The first set of above formula is closed since for every $t > 0$ $\eta(t,X)$ is closed.

If we set $A_t = \eta(t,\cdot)$, then

$$\{x \in X \mid x\cdot[0,T] \subset X\} = \bigcap_{t\in[0,T]} A_t^{-1}(X).$$

So also this set is closed. Therefore $G^T(X)$ is closed.

(vi) $G^T(X) \in \Sigma$ by (iv). $\eta(t,X) \in \Sigma$ by the continuity of η.

(vii) Let $\{x_n\} \subset \Gamma^T(X)$ with $x_n \to \bar{x}$. Then there exists $t_n \in [0,T]$ such that $x_n \cdot t_n \in \partial X$. Let t_n' be a subsequence of t_n converging to some $\bar{t} \in [0,T]$; then $x_n \cdot t_n' \to \bar{x}\cdot\bar{t} \in \partial X$. Therefore $\bar{x} \in \Gamma^T(X)$.

(viii) Let $x \in \Gamma^T(X)$; then $\exists\ t \in [0,T]$ such that $x \cdot t \in \partial X$; thus there exists a sequence $y_n \in X^c$ (X^c denotes the complement of X in M) converging to $x \cdot t$. This implies that $y_n(-t) \to x$. But $y_n(-t) \in G^T(X)$, therefore $x \in \partial G^T(X)$.

Now we can prove (1.1) a).

<u>Theorem 1.4</u> (Existence of an index pair). Let $X \in \Sigma$ and let T be large enough that $G^T(X) \subset \overset{\circ}{X}$. Then

$$(G^T(X),\ \Gamma^T(X))$$

is an index pair for X.

Proof. By lemma 1.3 (vi),(vii), $G^T(X)$ and $\Gamma^T(X)$ are closed. We have to check points (i),(ii) and (iii) of Def. 1.2.

(i) By lemma 1.2 (viii), $\overline{G^T(X) - \Gamma^T(X)} = G^T(X)$; so by lemma 1.2 (v), the conclusion follows.

(ii) Let $x \in \Gamma^T(X)$ and suppose that

(1-4) $\quad x\,[0,t] \subset G^T(X)$

we want to prove that $X \cdot [0,t] \subset \Gamma^T(X)$. Suppose that this fact is not true; then there exists $\bar{t} \in [0,t]$ such that $x \cdot \bar{t} \notin \Gamma^T(X)$.

Now set

$$t^* = \inf \{\tau \in [0,t] \mid x \cdot \tau \notin \Gamma^T(X)\}$$

Clearly $t^* \in [0,t)$ and

(1-5)

(a) $x \cdot t^* \in \Gamma^T(X)$ since $\Gamma^T(X)$ is closed by lemma 1.2 (vii);

(b) $x \cdot (t^* + \varepsilon_n) \notin \Gamma^T(X)$ (with $\varepsilon_n > 0$ and $\varepsilon_n \to 0$).

If we set $y = x \cdot t^*$, by (1.5) and the definition of $\Gamma^T(X)$ we have

$$y \cdot [0,T] \cap \partial X \neq \Phi$$
$$y \cdot [\varepsilon_n, T] \cap \partial X = \Phi$$

From the above formulas we have that

(1-6) $\quad y \in \partial X.$

On the other hand, by (1-4), $y \in G^T(X)$ and by our assumptions $y \in \overset{\circ}{X}$; this fact contradicts (1.6).

(iii) It is trivial

Theorem 1.5 (Equivalence of index pairs). Let $(N,N_\circ)$ and $(\tilde{N},\tilde{N}_\circ)$ be two index pairs such that exists $T > 0$ such that

$$G^T(\overline{N - N_\circ}) \subset \overline{\tilde{N} - \tilde{N}_\circ} \qquad \text{and}$$

$$G^T(\overline{\tilde{N} - \tilde{N}_\circ}) \subset \overline{N - N_\circ}$$

Then $[N/N_\circ] = [\tilde{N}/\tilde{N}_\circ]$.

Remark. The proof of theorem 1.5 is essentially contained in Salamon [S]. He gave a short and elegant proof of Conley's theorem of equivalence of index pairs (in the compact contest). Salamon's proof can be adapted to our case.

Sketch of the proof of th. 1.5. We can suppose that $G^T(\overline{N-N_\circ}) \subset \text{int } \tilde{N}-\tilde{N}_\circ$ and that $G^T(\overline{\tilde{N}-\tilde{N}_\circ}) \subset \text{int}(N-N_\circ)$ (if not it is enough to replace T by 2T and use lemma 1.2 (iv). Now let $f : N/N_\circ \to \tilde{N}/\tilde{N}_\circ$ be defined as follows

$$f([x]) = \begin{cases} [x \cdot 3T] & \text{if } x \cdot [0,2T] \subset N_1 - N_\circ \text{ or } x[T,3T] \subset \tilde{N}_1 - \tilde{N}_\circ \\ [N_\circ] & \text{otherwise} \end{cases}$$

the function f is continuous (for details of the proof see [S] lemma 4.7). In an analogous way we can define a map $\tilde{f} : \tilde{N}/\tilde{N}_\circ x[T,\infty) \to N/N_\circ$.

We have to prove that $\tilde{f} \circ f$ and $f \circ \tilde{f}$ are homotopic to the identity in N/N_o and $\tilde{N}/\tilde{N}_o$ respectively.

For $t \in [0,T]$ define the map $h : [0,T] \times N/N_o \to N/N_o$ as follows

$$h(t,[x]) = \begin{cases} [x \cdot 6t] & \text{if } x[0,6t] \subset N_1 - N_o \\ [N_o] & \text{otherwise.} \end{cases}$$

It is easy to show that h is continuous and that

$$h(T,[x]) = \tilde{f} \circ f \quad \text{and} \quad h(0,[x]) = Id_{N/N_o} .$$

In the same way it is possible to construct a homotopy $\tilde{h} : [0,T] \times \tilde{N}/\tilde{N}_o \to \tilde{N}/\tilde{N}_o$ .//

Corollary 1.6. If (N,N_o) and $(\tilde{N},\tilde{N}_o)$ are two index pairs for X, then $[N,N_o] = [\tilde{N},\tilde{N}_o]$. In particular (1-1) (b) holds.

Proof. If (N,N_o) and $(\tilde{N},\tilde{N}_o)$ are two index pairs for X, we have

$$G^T(\overline{N - N_o}) \subset G^T(X) \subset \tilde{N} - \tilde{N}_o \quad \text{by definition 1.1}$$

and

$$G^T(\overline{\tilde{N} - \tilde{N}_o}) \subset G^T(X) \subset N - N_o \quad .$$

The conclusion follows from theorem 1.5.//

So at this point h(X) is well defined. Another consequence of theorem 1.5 is the following Corollary

Corollary 1.7. Let $X,Y \in \Sigma$ and suppose that $\exists\, T \geq 0$ such that

(1-7) $\quad G^T(X) \subset Y \quad$ and $\quad G^T(Y) \subset X$

Then h(X) = h(Y).

Proof. Let $(N,N_\circ)$ and $(\tilde{N},\tilde{N}_\circ)$ be two index pairs for X and Y respectively. Then

(1-9) $\quad G^T(N-N_\circ) \subset G^T(X) \subset Y$ by definition 1.1 and (1-7).

Since $(\tilde{N},\tilde{N}_\circ)$ is an index pair for Y, $\exists\, T_1 > 0$ such that

$$G^{T_1}(Y) \subset \operatorname{int}(\tilde{N} - \tilde{N}_\circ)$$

Therefore by the above formula, (1-9) and lemma 1.2 (iii), we have that

$$G^{T+T_1}\ \overline{(N-N_\circ)} \subset \tilde{N} - \tilde{N}_\circ$$

For the same reason there exists $T_2 > 0$ such that

$$G^{T+T_2'} \subset \overline{(\tilde{N}-\tilde{N}_\circ)} \subset N - N_\circ$$

Thus by theorem 1.5 (replacing T with $T + \max(T_1, T_2)$) the conclusion follows.//

Corollary 1.8. For every $T > 0$ $\quad h(G^T(X)) = h(X)$.

Proof. Trivial.//

Corollary 1.9. If there is $T > 0$ such that $G^T(X) = \Phi$, then $h(X) = \underline{0}$.

Notice that Corollary 1.9 can be inverted as the following example shows.

Example 1.10. Take

$$M = \mathbb{R} \quad ; \quad \eta(t,x) = x-t \quad ; \quad X = [0,+\infty)$$

Then $h(X) = 1$ but $G^T(X) \neq \Phi$ for every $T > 0$.

However there is a good test to see if the index of a set is $\underline{0}$.

Theorem 1.11. Suppose that $X \in \Sigma$ and that

(1-10) for every $x \in X$, there is $t > 0$ such that $x \cdot t \in X$. Then $h(x) = \underline{0}$.

We need some lemmas to prove theorem 1.11.

Lemma 1.12. Suppose that (N,N_o) is an index pair and that τ is a positive constant such that

(1-11) $x \cdot [0,\tau] \subset N - N_o$

Then there exists an open neighbourhood V of x such that for every $y \in V \cap N$,

$$y \cdot [0,\tau] \subset N - N_o \quad .$$

Proof. We argue indirectly and suppose that the conclusion of the lemma is not true. Then exists a sequence $x_n \to X(x_n \in N - N_o)$ and a sequence $t_n \in [0,\tau]$ such that

$$x_n \cdot t_n \notin N - N_o \ .$$

We set

$$\tilde{t}_n = \sup \{t \in [0,t_n] \text{ such that } x \cdot [0,t] \subset N\}$$

$\tilde{t}_n$ is a bounded sequence; so we can suppose that it is convergent to some $\bar{t} \in [0,\tau]$. By our construction, $x_n \cdot \tilde{t}_n \in N_\circ$; so $x \cdot \bar{t} \in N_\circ$ since $N_\circ$ is closed. This last statement contradicts (1-11); so the lemma is proved.//

Lemma 1.13. Let $(N,N_\circ) = (G^T(X),\Gamma^T(X))$ be an index pair for X (cf. Th. 1.4). We set.

$$U = \{x \in N \mid \exists\, t \in [0,2T] \text{ such that } x \cdot t \quad N^C\}$$

where N^C denotes M - N.

Then U satisfies the following properties

(i) U is relatively open in N;

(ii) given two positive constants $t_1 < t_2$ such that

$$x \cdot t_i \in U \quad \text{and} \quad X \cdot [0,t_i] \subset N \quad (i=1,2)$$

then for every $t \in [t_1,t_2]$, $x \cdot t \in U$;

(iii) $N_\circ \subset U$;

(iv) $(\bar{U},N_\circ)$ is an index pair and $[\bar{U}/N_\circ] = \underline{0}$.

Proof. (i) and (ii) are easy to check.

In order to prove (iii) we argue indirectly and suppose that there is $x \in N_\circ$ such that $x \cdot [0,2T] \subset N$. Since $N_\circ$ is positively invariant with respect to N, $x \cdot [0,2T] \subset N_\circ$. Then if we set $y = x \cdot T$, it follows that $y \in N_\circ$ and $y \in G^T(N)$. Since $G^T(N) \subset \overset{\circ}{N}$ by lemma 1.3 (iii) and (iv) and $N_\circ \subset \partial N$, by lemma 1.3 (viii), we have obtained a contradiction.

Now let us prove (iv). First observe that $N_\circ \subset \overline{U}$ by (iii). (i) of Def. 1.1 is satisfied since $\overline{\overline{U} - N_\circ} = \overline{U}$ and $G^{2T}(\overline{U}) = \Phi \subset \text{int}(\overline{U})$.

To check (ii), it is enough to observe that $\overline{U} \subset N$. (iii) follows directly by the definition of $\overline{U}$. So $(\overline{U}, N_\circ)$ is an index pair.

$$[\overline{U}/N_\circ] = h(U) = \underline{0} \text{ by Corollary (1-9).//}$$

<u>Proof of Th. 1.11</u>. Let N, $N_\circ$ and U as in lemma 1.13. For every $x \in N$, we choose a $t(x) > 0$ such that

$$x\cdot[0,t(x)] \subset N \quad \text{and} \quad x\cdot t(x) \in U .$$

This is possible by (1-10) and lemma 1.13 (iii). If $x \in U$ we choose $t(x) = 0$. Also if $x \notin U$, we can choose $t(x)$ such that $t(x) \notin N_\circ$.

Now for $x \in N - N_\circ$, let V_x be an open neighbourhood of N (open in the topology of N) such that

(1-12) for every $y \in V_x$, $y\cdot[0,t(x)] \subset N$ and $y\cdot t(X) \in U$.

This is possible by our choice of t(x), lemma 1.12 and lemma 1.13 (i).

For $x \in N_\circ$, set $V = U$. Thus $\{V_x\}_{x \in N}$ is an open cover of N (open in the relative topology of N).

Let $\{V_i\}_{i \in I}$ be a locally finite refinement of $\{V_x\}_{x \in N}$ which exists since N is a metric space.

Observe that, by our construction, for every $i \in I$, there exists $t_i \geqq 0$ such that

(1-13) $\eta(t_i, V_i) \subset U$ and $\eta([0,t_i], V_i) \subset N$.

Now let $\{\beta_i(x)\}_{i \in I}$ be a partition of the unity relative to $\{V_i\}_{i \in I}$ i.e. a set of function $\beta_i : N \to \mathbb{R}$ whose support is $\overline{V}_i$ and $\sum_{i \in I} \beta_i(X) = 1$ for every $x \in N$. Such partition exists since N is a metric space.

Now set

$$\tau(x) = \sum_{i \in I} \beta_i(x) t_i .$$

Clearly $\tau(x)$ is a continuous function. We claim that

(1-14) $x \cdot \tau(x) \in U$.

In order to see this, fix $\overline{x} \in N$ and set

$$t_1(\overline{x}) = \min\{t_i \mid \overline{x} \in V_i\} \quad ; \quad t_2(\overline{x}) = \max\{t_i \mid \overline{x} \in V_i\} .$$

By (1-13), $\eta(t_i, \overline{x}) \in U$ (i=1,2) and $\eta([0,t_1], \overline{x}) \subset N$. Therefore (1.14) follows from lemma 1.13 (ii).

Moreover observe that by our construction

(1-15) $\tau(x) = 0$ for every $x \in N_o$.

Now consider the map $h : [0,1] \times N \to U$ defined by

$$h(s,x) = \eta(s \cdot t(x), x)$$

h is an homotopy equivalence between N and $\overline{U}$, and by (1-15) it is also an homotopy equivalence between N/N_o and $\overline{U}/N_o$.

Therefore, by lemma 1.13 (iv)

$$h(X) = [N/N_\circ] = [\overline{U}/N_\circ] = \underline{0} \ . \ //$$

Remark 1.14. Now, few words to compare the Conley index with our generalization.

A closed set X is called by Conley [C] an insolated neighbourhood if $I(X) \subset \overset{\circ}{X}$ where $I(X) = \{x \in X : x\,\mathbb{R} \subset X\}$ or, using our notation, $I(X) = \bigcap_{t \geq 0} G^T(X)$.

Let $\tilde{\Sigma}$ be the family of isolating neighbourhoods in M; then if M is compact $\Sigma = \tilde{\Sigma}$. If M is not compact, in general, $\Sigma \subset \tilde{\Sigma}$. So, in our approach, it was necessary to restrict the class of sets X for which to define index pairs (and introduce the operator $G^T(\cdot)$).

Now, observe that the relationship (1-7) gives an equivalence relation on Σ (which we will denote by $\approx$).

Corollary 2.4 states that the index is constant on each equivalence class of $\approx$. If M is compact, then $X \approx Y$ if and only if $I(X)=I(Y)$ (the easy proof of this is left to the reader).

So, when M is compact, h depends only on the maximal invariant set I(X) contained in X; therefore it is an index of isolated invariant sets. Example 1.10 shows that this is not the case when the compactness is not assumed (in fact $h(X) = \underline{1}$ but $I(X) = \Phi$). Concluding the Conley index is an index of isolated invariant sets; our generalization is an index of a class of closed set Σ which has been closed in order that the main properties of the Conley theory can be preserved.

Example 1.15. Let M = E be an Hilbert space and let L be a bounded normal invertible operator.

We consider the flow η defined by the differential equation

(1-16) $\dot{x} = L\,x$

We want to compute $h(X,\eta)$ where X is a bounded closed neighbourhood of 0. By our assumption E can be splitted as follows

(1-17) $E = E^+ \oplus E^-$

where E^+ and E^- are two mutually orthogonal subspaces such that exists a constant $\alpha > 0$

$$\text{(1-18)}\quad \begin{aligned} &\langle L\,x, x\rangle \geqq \alpha \| x \|^2 \qquad \forall\, x \in E^+ \\ &\langle L\,x, x\rangle \leqq -\alpha \| x \|^2 \qquad \forall\, x \in E^- \end{aligned}$$

According to the splitting (1-17), (1-16) can be written as follows

$$\begin{aligned} \dot{x}^+ &= L^+ x^+ \\ \dot{x}^- &= L^- x^- \end{aligned}$$

where $x = x^+ + x^-$ with $x^\pm \in E^\pm$ and $L^\pm = L|_{E^\pm}$.

Now, if Y is any other bounded closed neighbourhood of 0, by 1-18 it is easy to check that $X, Y \in \Sigma(\eta)$ and that (1-7) is satisfied. Then $h(X) = h(Y)$. In particular we can take

$$Y = (B_R \cap E^+) \times (B_R \cap E^-)$$

where B_R is the ball of radius R.

It is easy to check that

$$(Y, (B_R \cap E^+) \times \partial(B_R \cap E^-))$$

is an index pair and that it is homotopically equivalent to

$$(B_R \cap E^-, \partial(B \cap E^-)) .$$

Also we have

$$\begin{bmatrix} B_R \cap E^- \\ \partial(B_R \cap E^-) \end{bmatrix} = \begin{cases} [S^N,*] & \text{if } \dim E^- = N \\ [S^\infty,*] = \underline{0} & \text{if } E^- \text{ is infinite dimensional} \end{cases} .$$

So concluding we have

$$h(X) = h(Y) = [S^N,*]$$

where N is dim E^- and remembering that $[S^\infty,*] = [*,*] = \underline{0}$.

2. STABILITY AND HOMOTOPY INVARIANCE OF THE GENERALIZED CONLEY INDEX

We need new notation:

$$\mathcal{F}(M) = \{X \subset M \mid X \text{ is closed}\}$$

$\mathcal{F}(M)$ can be equipped with the Hausdorff metric:

$$d_H(X,Y) = \sup_{x \in X} d(x,Y) + \sup_{y \in Y} d(y,X)$$

We need also the following notation

$$X \overset{\delta}{\subset} Y \overset{\text{def}}{\Longleftrightarrow} X \subset Y \quad \text{and} \quad \operatorname{dist}(X, \partial Y) \geqq \delta$$

$$X \subset Y \overset{\text{def}}{\Longleftrightarrow} \exists \delta > 0 \ \text{ s.t. } \ X \overset{\delta}{\subset} Y .$$

We set

$$\Sigma_o = \Sigma_o(\eta) = \{X \in \mathcal{F}(M) \mid \exists\, T,\delta > 0 \text{ s.t. } G^T(N_\delta(X)) \subset X, \text{ and } \eta(\cdot,T) \text{ is uniformly continuous}\} .$$

Clearly $\Sigma_o \subset \Sigma$.

Theorem 2.1. Let $X \in \Sigma_o(\eta)$ and let $\tilde{\eta}$ be a flow such that

$$d(\eta(t,x), \tilde{\eta}(t,x)) \leqq \varepsilon \qquad \forall\, t \quad [-T,T], \ \forall\, x \in X$$

where ε and t are suitable positive constants, which depend on X and η. Then,

(i) $X \in \Sigma_o(\tilde{\eta})$

(ii) $h(X,\tilde{\eta}) = h(X,\eta)$

Before proving theorem 2.1 we will see two important consequences of this theorem.

Corollary 2.2. Let $X, \eta, \tilde{\eta}$ be as in theorem 2.1 and let $\tilde{X}$ be a closed set such that

(2-1) $\qquad d_H(X,\tilde{X}) < \varepsilon_1$

where $d_H(\cdot\,,\cdot)$ is the Hausdorff distance and ε_1 is a positive distance depending on $X,\eta,\tilde{\eta}$ but not on $\tilde{X}$.

Then

(i) $X \in \Sigma_o(\tilde{\eta})$

(ii) $h(\tilde{X},\tilde{\eta}) = h(X,\eta)$.

Proof. By Th. 2.1 (i), there exists $T,\delta_1,\delta_2 > 0$ such that

$$G^T(N_{\delta_1}(X),\tilde{\eta}) \overset{\delta_2}{\subset} X$$

Now choose $\varepsilon_1 > 0$ smaller than min $(\delta_1/2, \delta_2/2)$. Then by (2-1),

$$\text{(2-2)} \qquad G^T(N_{\delta_1}(X),\tilde{\eta}) \overset{\varepsilon_1}{\subset} \tilde{X} \cap X$$

Moreover, by the choice of ε_1, we have

$$\tilde{X} \subset N_{\varepsilon_1}(X) \subset N_{\delta_1/2}(X) \quad \text{and} \quad N_{\varepsilon_1}(\tilde{X}) \subset N_{\delta_1}(X) .$$

By the above formula and (2-2) we get

$$\text{(2-3)} \qquad G^T(N_{\varepsilon_1}(\tilde{X}),\tilde{\eta}) \subset G^T(N_{\delta_1}(X),\tilde{\eta}) \overset{\varepsilon_1}{\subset} X \cap \tilde{X} .$$

The above formula proves (i).

Moreover by (2-2) and (2-3), we have that

$$G^T(X,\tilde{\eta}) \subset \tilde{X} \quad \text{and} \quad G^T(\tilde{X},\tilde{\eta}) \subset X .$$

Then by corollary 1.7 we have

$$h(X,\tilde{\eta}) = h(\tilde{X},\tilde{\eta}) .$$

The conclusion follows by (ii) of Theorem 2.1.//

Corollary 2.3. Let $\eta_\lambda, \lambda \in [0,1]$, be a family of flows depending continuously on λ with respect to the topology of the uniform convergence on $X \times [-T,T]$ for every $T > 0$ where $X \subset M$.

Suppose that X_λ is a family of sets contained in X and depending uniformly on λ with respect to the Hausdorff topology.

Finally suppose that $X_\lambda \in \Sigma_o(\eta_\lambda)$ for every $\lambda \in [0,1]$.

Then $h(X_\lambda, \eta_\lambda)$ does not depend on λ.

Proof. By Corollary 2.2, for every $\bar{\lambda} \in [0,1]$, there exists a neighbourhood of $\bar{\lambda}$, $I_{\bar{\lambda}}$ such that

$$h(X_\lambda, \eta_\lambda) \quad \text{is constant for } \lambda \in I_{\bar{\lambda}} .$$

Then the conclusion follows straightforwardly.

The proof of theorem 2.1 is involved and relies on several lemmas

Lemma 2.4. Take $X \in \Sigma$ and T large enough such that

$$\text{(2-4)} \qquad G^{T/2}(X) \subset \mathring{X}$$

Set Φ_1: $\text{int}(G^{T/2}(X) \to G^T(X)/\Gamma^T(X)$ be defined as follows

$$\Phi_1(x) = \begin{cases} [x \cdot T] & \text{if } x \cdot T \in G^T(X) \\ [\Gamma^T(X)] & \text{if } x \cdot T \notin G^T(X) . \end{cases}$$

Then Φ_1 is continuous.

Proof. It is obvious that $\Phi_1(x)$ is continuous if $x \cdot T \in \text{int}(G^T(x))$ or $x \cdot T \notin G^T(X)$. So we have to consider only the case $x \cdot T \in \partial G^T(X)$. First notice that

$$\text{(2-5)} \qquad x \in \text{int}(G^{T/2}(X)) \Longrightarrow X \cdot [0,T/2] \subset \mathring{X} .$$

Moreover

$$x \cdot T \in G^T(X) \Longrightarrow x \cdot [\frac{1}{2} T, \frac{3}{2} T] \subset G^{T/2}(2)$$

Thus by (2-4) and the above formula $x[\frac{1}{2} T, \frac{3}{2} T] \subset \overset{\circ}{X}$ and by (2-5) it follows that

(2-6) $\quad x[0, \frac{3}{2} T] \subset \overset{\circ}{X}$.

We claim that

(2-7) $\quad x \cdot T \in \partial G^T(X) \Longrightarrow x \cdot T \in \Gamma^T(X)$.

In fact if $x \cdot T \in \partial G^T(X)$ there exists $t \in [0,2T]$ such that $x \cdot t \in \partial X$.

By (2-6) we have that $t \geqq \frac{3}{2} T \geqq T$. Then by definition of $\Gamma^T(X)$, (2-6) follows. So we have that

$$x \cdot T \in \partial G^T(X) \Longrightarrow \Phi_1(x) = [\Gamma^T(X)]$$

and by the above formula the continuity of Φ_1 at x follows easily.//

Lemma 2.5. The function $\Phi_2 : G^T(X)/\Gamma^T(X) \to G^T(X)/\Gamma^T(X)$ defined as follows

$$\Phi_2([x]) = \begin{cases} [x \cdot T] & \text{if } x \cdot T \in G^T(X) \\ [\Gamma^T(X)] & \text{otherwise} \end{cases}$$

is continuous.

Proof. The proof of this lemma is contained in the proof of Th. 1.5 when it is shown that $h(t,[x])$ is continuous.//

Lemma 2.6. Let $X \in \Sigma_o(\eta)$ and let $T > 0$ be large enough that

(2-8) $\quad G^{T/2}(X) \overset{\delta}{\subset} X \qquad$ for some $\delta > 0$

Then there exists $\delta_1 = \delta_1(\eta, X)$ such that

$$x \in N_{\delta_1}(\Gamma^T(X)) \implies x\cdot[0, \tfrac{1}{2}T] \cap \partial X \neq \phi .$$

Proof. Choose δ_1 small enough that

(2-9) $\quad d(x_1, x_2) < \delta_1 \implies d(x_1\cdot T, x_2\cdot T) \leqq \delta/2 \qquad \forall\, x_1, x_2 \in X .$

This is possible by the uniform continuity of $\eta(T,\cdot)$.

So we have

$x \in N_{\delta_1}(\Gamma^T(X)) \implies$

$\exists\, \bar{x} \in \Gamma^T(X) : d(x,\bar{x}) \leqq \delta_1 \implies$ [by (2-9)]

$d(x\cdot T, \bar{x}\cdot T) < \delta/2 \implies$ [since $\bar{x}\cdot T \in \partial X$]

$d(x\cdot T, \partial X) < \delta/2 \implies$ [by (2-8)]

$x\cdot T \not\in G^{T/2}(X) \implies$ [by the definition of $G^{T/2}(X)$]

$x\cdot[\tfrac{1}{2}T, \tfrac{3}{2}T] \cap \partial X \neq \phi$. //

In the following lemmas we shall write $\eta_t(x)$ instead of $\eta(t,x)$ to simplify the notation.

Lemma 2.7. Take $X \in \Sigma_o(\eta)$ and choose T large enough that

(2-10) $\quad G^T(X) \overset{\delta}{\subset} G^{T/2}(X) \overset{\delta}{\subset} \overset{\circ}{X}$.

Let $\tilde{\eta}$ be a flow such that

$$(2\text{-}11) \qquad d(\tilde{\eta}_t(x),\eta_t(x)) \leqq \frac{\delta_1}{2} \qquad \forall\, x \in X \qquad \forall\, t \in [-T,T]$$

where $\delta_1 = \delta_1(\eta,X) < \delta$ is defined in lemma 2.6.
Let $h : [0;1] \times G^T(X)/\Gamma^T(X) \to G^T/\Gamma^T(X)$ be defined as follows

$$h(\lambda,[x]) = \begin{cases} [\eta_{2T} \circ \tilde{\eta}_{-\lambda T} \circ \eta_{\lambda T}(x)] & \text{if } \eta_{[0,2T]}(x) \subset G^T(X) \\ [\Gamma^T(X)] & \text{otherwise} \end{cases}$$

Then h is continuous.

Proof. By (2-11) taking $t = -\lambda T$ and replacing x with $\eta_{-\lambda T}(x)$ we get

$$(2\text{-}12) \qquad d(\tilde{\eta}_{-\lambda T} \circ \eta_{\lambda T}(x),x) \leqq \delta_1/2 \ (\leqq \delta/2) \quad \forall\, \lambda \in [0,1] \quad \forall\, x \in G^T(X).$$

Then by (2-10), the function

$$x \to \tilde{\eta}_{-\lambda T} \circ \eta_{\lambda T}(x)$$

maps $G^T(X)$ into $\text{int}(G^{T/2}(X))$ for every $\lambda \in [0,1]$.

Now consider the function $g : [0,1] \times G^T(X) \to G^T(X)/\Gamma^T(X)$ defined as follows

$$g(\lambda,x) = \begin{cases} [\eta_{2T} \circ \eta_{-\lambda T} \circ \eta_{\lambda T}(x)] & \text{if } \eta_{[0,2T]} \subset G^T(X) \\ [\Gamma^T(X)] & \text{otherwise .} \end{cases}$$

Then we have $g(\lambda,x) = \phi_2 \circ \phi_1 \circ (\tilde{\eta}_{-\lambda T} \circ \eta_{\lambda T})$ where

$$\tilde{\eta}_{-\lambda T} \circ \eta_{\lambda T} : [0,1] \times G^{T}(X) \to G^{T/4}(X)$$

$\phi_1 \qquad : G^{T/4}(X) \to G^{T}(X)/\Gamma^{T}(X)$ is defined by lemma 2.4

$\phi_2 \qquad : G^{T}(X)/\Gamma^{T}(X) \to G^{T}(X)/\Gamma^{T}(X)$ is defined by lemma 2.5 .

Since all the above maps are continuous also g is continuous.

It remains to prove that

$$h(\lambda,[x]) = g(\lambda,x) .$$

So we have to prove that if $x \in \Gamma^{T}(X)$ then $g(t,x)$ is constant, so that the above equality makes sense.

By (2-12) we have

$$x \in \Gamma^{T}(X) \Longrightarrow \tilde{\eta}_{-\lambda T} \circ \nu_{\lambda T}(x) \in N_{\delta_1}(\Gamma^{T}(X)) .$$

By the above formula and lemma 2.6 we have that

$$x \in \Gamma^{T}(X) \Longrightarrow \eta_{[0,2T]} \circ \eta_{\lambda T} \circ \eta_{-\lambda T}(x) \cap \partial X \neq \phi .$$

Therefore $g(\lambda,x) = [\Gamma^{T}(X)] \quad \forall\, x \in \Gamma^{T}(X)$. //

Lemma 2.8. Take T large enough that

$$G^{T}(X) \overset{\delta}{\subset} G^{T/2}(X) \overset{\delta}{\subset} X$$

and take $\delta_1 = \delta_1(X,\eta) < \delta/4$.

Now take $\tilde{\eta}$ close enough to η such that

$$\text{(i)} \quad d(\tilde{\eta}_t(x), \eta_t(x)) < \delta_1 \quad \text{for every } x \in X \text{ and } t \in [-2T, 2T]$$

$$\text{(2-13)} \quad \text{(ii)} \quad N_{\delta_1}(G^T(X)) \subset \tilde{G}^{T/2}(X)$$

$$\text{(iii)} \quad \tilde{G}^T(X) \overset{2\delta_1}{\subset} X$$

Then the function $f : G^T(X)/\Gamma^T(X) \to \tilde{G}^T(X)/\tilde{\Gamma}^T(X)$ defined as follows

$$f([x]) = \begin{cases} [\tilde{\eta}_{2T} \circ \eta_{-T}(x)] & \text{if } \tilde{\eta}_{2T} \circ \eta_T(x) \in \tilde{G}^T(x) \\ [\tilde{\Gamma}^T(x)] & \text{otherwise} \end{cases}$$

is continuous (we have used the notation $\tilde{G}^T(X) = G^T(X,\tilde{\eta})$ and $\tilde{\Gamma}^T(X) = \Gamma^T(X,\tilde{\eta})$).

<u>Proof</u>. By (2-13) (i) we set

$$\text{(2-14)} \quad d(\tilde{\eta}_{-t} \circ \eta_t(x), x) < \delta_1 \quad \forall\, x \in G^T(X) \quad \forall\, t \in [0,T].$$

Then, by (2.13) (ii), the function $\tilde{\eta}_T \circ \eta_T$ maps $G^T(X)$ into int $G^{T/2}(X)$. Now define $g : G^T(X) \to \tilde{G}^T(X)/\tilde{\Gamma}^T(X)$ as follows

$$g(x) = \begin{cases} \tilde{\eta}_{2T} \circ \eta_{-T}(x) & \text{if } \tilde{\eta}_{2T} \circ \eta_{-T}(x) \in \tilde{G}^T(X) \\ [\tilde{\Gamma}^T(x)] & \text{otherwise .} \end{cases}$$

Notice that

$$g(x) = \tilde{\phi}_1 \circ (\tilde{\eta}_T \circ \eta_{-T})$$

where $\tilde{\phi}_1 : \text{int}\, \tilde{G}^{T/2}(X) \to \tilde{G}^T(X)/\tilde{\Gamma}^T(X)$
is the map of lemma 2.4 with $G^T(X)$, $\Gamma^T(X)$ and η_t replaced by $\tilde{G}^T(X)$, $\tilde{\Gamma}^T(X)$, and $\tilde{\eta}_t$ respectively.

Therefore g is continuous.

It remains to prove that

$$f([x]) = g(x) .$$

So we have to prove that

$$x \in \Gamma^T(X) \Longrightarrow g(x) \quad \text{is constant}$$

or more exactly $g(x) = [\tilde{\Gamma}(X)]$.

Use (2-13) (i) with $t = 2T$ and x replaced by $\eta_{-T}(x)$ with $x \in \Gamma^T(X)$ then we have

$$d(\tilde{\eta}_{2T} \circ \nu_{-T}(x), \eta_{2T} \circ \eta_{-T}(x)) \leqq \delta_1$$

or

$$d(\tilde{\eta}_{2T} \circ \eta_{-T}(x), \eta_T(x)) < \delta_1$$

Since $x \in \Gamma^T(X)$, we have that $\eta_T(X) \in \partial X$, and by the above formula

$$d(\tilde{\eta}_{2T} \circ \eta_{-T}(x), \partial X) \leqq \delta_1 .$$

Thus $\tilde{\eta}_{2T} \circ \nu_{-T}(x) \notin \tilde{G}^T(X)$. So we have proved that

$$x \in \Gamma^T(X) \Longrightarrow g(x) = \tilde{\Gamma}^T(X)$$

and this completes the proof of the lemma.//

<u>Proof of Theorem 2.1</u>. Take T and ε such that (2-11) and (2-13) are satisfied with $\delta_1 < 2\varepsilon$.

Moreover, if ε is small enough, we have also

(2-14)
(a) $N_{\delta_1}(\tilde{G}^T(X)) \subset G^{T/2}(X)$

(b) $\tilde{G}^T(X) \overset{2\delta_1}{\subset} X$

Now let $f : G^T(X)/\Gamma^T(X) \to \tilde{G}^T(X)/\tilde{\Gamma}^T(X)$ be the function defined in lemma 2.8. We have to prove that f is an homotopy equivalence.

We claim that $\tilde{f} : \tilde{G}^T(X)/\tilde{\Gamma}^T(X) \to G^T/\Gamma^T(X)$ is the homotopy inverse of f ($\tilde{f}$ is defined as f replacing $G^T(X)$ with $\tilde{G}^T(X)$, etc. ...). f and $\tilde{f}$ are continuous by virtue of lemma 2.8 and (2.14). Moreover $\tilde{f} \cdot f =$ $= h(1,\cdot)$ where h is defined in lemma 2.7.

Lemma 2.7 shows that $\tilde{f} \circ f \sim h(0,\cdot)$ (where "$\sim$" means homotopy equivalence). Moreover it is straightforward to show that $h(0,\cdot) \sim \mathrm{Id}$. Thus $\tilde{f} \circ f \sim \mathrm{Id}$. Analogously we can show that $f \circ \tilde{f} \sim \mathrm{Id}$ and this proves Theorem 2.1.

<u>Example 2.9</u>. Let η be the flow defined on M by differential equation

$$\dot{x} = F(x) .$$

We suppose that M is an Hilbert space E (or an Hilbert manifold). Let $\bar{x}$ a nondegenerate critical point for F i.e. $F(\bar{x}) = 0$ and $F'(\bar{x}) :$ $: T_{\bar{x}} M \to T_{\bar{x}} M$ (where $T_{X_o} M$ denotes the tangent space at X_o) is defined (as Frechêt derivative) and it is an invertible normal operator.

Since $F'(x)$ is a normal operator, we have (cf. Ex. 1.15)

$$T_{\bar{x}} M = E^+ \oplus E^-$$

where E^+ is the stable manifold of η and E^- the unstable manifold.

Now let η_o be the flow defined by the following equation

$$\dot{x} = \overline{x} + F'(x_o) \cdot x \, .$$

By Theorem 2.1 it follows that

$$h(U,\eta) = h(U,\eta_o)$$

where U is a neighbourhood of $\overline{x}$ sufficiently small.

Therefore by example 1.15, it follows that

$$(2\text{-}15) \qquad h(U,\eta) = (S^{m(x)},*)$$

where $m(x) = \dim E^-$.

3. THE GENERALIZED CONLEY INDEX AND COMPACTNESS.

For $X \in \Sigma(\eta)$ we set

$$I(X) = \bigcap_{T>0} G^T(X) = \{x \in X \mid \eta(t,x) \in X \text{ for every } t \in \mathbb{R} \text{ such that } \eta(t,x) \text{ is defined}\}$$

The following compactness assumption is very important for our theory:

Def. 3.1. Let $X \in \Sigma$. We say that X satisfies the property (C) if for every neighbourhood U of I(X) there exists T > 0 such that

$$G^T(X) \subset U \, .$$

Observe that the property (C) is hereditary, i.e. if X satisfies the property (C) and $Y \subset X$ ($Y \in \Sigma$), then Y satisfies the property (C).

Prop. 3.2. Suppose that $X, Y \in \Sigma$ and satisfy the property (C). Then

$$I(X) = I(Y) \Longrightarrow h(X) = h(Y) .$$

Proof. Let $S = I(X) = I(Y)$. $U = X \cap Y$ is a neighbourhood of S. Then, since U and V satisfy the property (C) there exists $T > 0$ such that

$$G^T(X) \subset U \subset Y \quad \text{and} \quad G^T(Y) \subset U \subset X .$$

The conclusion follows by Corollary 1.7.//

Def. 3.3. We say that $S \subset X$ is a (C)-invariant set if

(i) S is an invariant set

(ii) S has a neighbourhood U which satisfies the property (C) and such that $I(U) = S$.

By the remarks before Prop. 3.3 and by Prop. 3.3, it follows that any neighbourhood sufficiently small of S has the same homotopy index. Therefore it is natural to define the index of a (C)-invariant set S as follows:

(3-1) $\quad h(S) = h(U) \quad$ where $U \in \Sigma$ neighbourhood of S sufficiently small.

The following proposition gives a criterium to check if a set $U \in \Sigma$ satisfies the property (C).

Prop. 3.4. Let $U \in \Sigma$ and suppose that

(3-2) given a sequence $x_n \in U$ and a sequence $t_n \to +\infty$ such that $x_n \cdot [0, t_n] \subset U$, then the sequence $x_n \cdot t_n$ has a limit point.

Then U satisfies the property (C).

Proof. We argue indirectly and suppose that there exists a neighbourhood V of I(U) such that for every T > 0

$$G^T(U) \not\subset V \ .$$

Then there exists a sequence $y_n \in U$ and a sequence $t_n \to +\infty$ such that

$$y_n \in G^{t_n}(X) - V \ .$$

If we set $x_n = y_n(-t_n)$, then $x_n \cdot [0,t_n] \subset U$. Then by (3-2) $x_n \cdot t_n$ is convergent to some $\bar{y}$ (may be considering a subsequence). By its construction $\bar{y} \cdot \mathbb{R} \subset U$, therefore $\bar{y} \in S$.

However, since $\bar{y} = \lim_{n \to +\infty} y_n$ we have that $\bar{y} \notin \mathring{V}$. And this is a contradiction since V is a neighbourhood of S.//

Corollary 3.5. Let M be a locally compact space. Then any compact invariant isolated set $S \subset M$ is a (C)-invariant set.

Therefore, the index (3-1) is defined for such S.

Proof. Clearly every compact neighbourhood of S satisfies (3-2).//

Remark 3.6. When M is locally compact we get the "classical" Conley theory (of Remark 1.14).

The property (3.2) (which was introduced by Rybakowsky [R]) can replace the local compactness of M in such a way that the main properties of the "original" Conley index are preserved (in particular it is possible to define the index of an isolated invariant set).

Our theory has been developed without any request of compactness, replacing the index of an invariant set with the index of a set $X \in \Sigma$.

A compactness property, as the property (C), is required only to define the index of an invariant set as in the original Conley theory.

Prop. 3.7. Let U satisfy the property (C) and suppose that I(U) is compact. Then $U \in \Sigma_o$.

Proof. Let

$$\varepsilon = d(\partial U, I(U)) .$$

Since I(U) is compact then $\varepsilon > 0$. Then setting $V = N_{\varepsilon/2}(I(U))$, we have that $V \in \Sigma$ and that, for T large enough

$$G^T(U) \subset V \quad \text{(since U satisfies the property (C))}.$$

Thus $V \overset{\varepsilon/2}{\subset} U$ as we wanted to prove.//

Example 3.8. Let $\bar{x}$ be as in example 2.9. Then $\bar{x}$ is a (C)-invariant set and

$$h(\bar{x}) = (S^{m(\bar{x})},*) .$$

4. THE GENERALIZED MORSE INDEX.

Let $H^*(\cdot,\cdot)$ denote the Alexander-Spanier cohomology with coefficients in some field F (cf. [Sp]).

We recall that the Alexander-Spanier cohomology satisfies the following property which is not shared by the singular cohomology theory.

Th. 4.1. Let (X,A) and (Y,B) two pairs of topological spaces. We suppose that X and Y are paracompact Hausdorff spaces and that A and B are closed in X and Y respectively. Moreover suppose that X - A and Y - B are homeomorphic. Then

$$H^*(X,A) \sim H^*(Y,B) .$$

Proof. See [Sp], Th. 5, pag 318.//

Now for every pair of closed spaces (X,A) we set

$$p(X,A) = p_t(X,A) = \sum_{q=0}^{\infty} [\dim H^q(X,A)]\, t^q$$

p(X,A) is a formal series whose coefficients are cardinal numbers; these numbers are known as Betti numbers.

If X is a compact manifold with boundary A, then p(X,A) reduces to a polynomium, called Poincaré or Betti polynomium.

p(X,A) is a topological invariant which carries part of the information contained in the cohomology algebra $H^*(X,A)$.

When $A = \Phi$ we shall write p(X) instead of $p(X,\Phi)$. We shall denote by S the set of formal series with cardinal coefficients.

The following properties of p(X,A) will be used to study the generalized Morse index.

Lemma 4.2. Let (X,A) and (Y,B) be couples of closed subspaces of a metric space. Then

(i) $p(X,A) = p(X/A,[A])$

(ii) if $X \cap Y = \Phi$ then
$p(X \cup Y, A \cup B) = p(X,A) + p(Y,B)$

(iii) $p((X,A) \times (Y,B)) = p(X,A) \cdot p(Y,B)$
where $(X,A) \times (Y,B) = (X \times Y, X \times B \cup Y \times A)$

(iv) if $B \subset A \subset X$ then there exists $Q(t) \in S$ s.t.
$p_t(X,A) + p_t(A,B) = p_t(X,B) + (1+t)Q(t)$.

Proof. (i) Let $\pi : X \to X/A$ be the projection map. Then $\pi_{|X-A}$ is a homeomorphism between X-A and X/A - [A]. Thus the conclusion follows from Th. 4.1.

(ii) trivial.

(iii) Since (X,A) and (Y,B) are closed pairs, there is an exact Mayer-Vietoris sequence for the $\bar{H}^*$ cohomology (cf. [Sp] pag 291).

But every closed pair of Hausdorff-paracompact spaces is a tout pair for the Alexander-Spanier cohomology (cf. [Sp] pag 315).

Therefore $\bar{H}^* = H^*$ on such pairs. Therefore the Künneth formula can be applied to such pairs (cf. [Sp] pag 249) and we get

$$H^*((X,A) \times (Y,B)) = H^*(X,A) \otimes H^*(Y,B) .$$

From the above formula the conclusion follows.

(iv) Let us consider the exact sequence relative to the triple $B \subset A \subset X$:

$$(4\text{-}1) \qquad \cdots \xrightarrow{\delta^*_{q-1}} H^q(X,A) \xrightarrow{i^*_q} H^q(X,B) \xrightarrow{j^*_q} H^q(B,C) \xrightarrow{\delta^*_q} \cdots$$

and set
$$a_q = \dim (\ker i^*_q)$$
$$b_q = \dim (\ker j^*_q)$$
$$c_q = \dim (\ker \delta^*_q)$$

By the exactness of (4-1) we get

$$\dim H^q(X,A) = c_{q-1} + a_q \quad \text{(with the convention that } c_{-1} = 0)$$
$$\dim H^q(X,B) = a_q + b_q$$
$$\dim H^q(A,B) = b_q + c_q$$

Then we have

$$p(X,A) = \sum_{q=0}^{\infty} (c_{q-1} + a_q)t^q$$

$$p(X,B) = \sum_{q=0}^{\infty} (a_q + b_q)t^q$$

$$p(A,B) = \sum_{q=0}^{\infty} (b_q + c_q)t^q$$

Then

$$p(X,A) + p(A,B) = p(X,B) + \sum_{q=0}^{\infty} (c_{q-1} + c_q)t^q = p(X,B) + (1+t)\sum_{q=0}^{\infty} c_q t^q$$

The conclusion follows setting $Q(t) = \sum_{q=0}^{\infty} c_q t^q$.

Notice that the formula (iv) holds even if some of the coefficients are infinite cardinal numbers.//

We can now define the generalized Morse index:

Def. 4.3. The generalized Morse index (GIM) is a map

$i : \Sigma(\eta) \to S$

defined by

$i_t(X,\eta) = p_t(N,N_\circ)$

where $(N,N_\circ)$ is an index pair for X.

When no ambiguity is possible we shall write i(X) instead of $i_t(X,\eta)$.

Using Th. 1.4 we could define the GIM in the following (formally) simpler way

$$i_t(X,\eta) = \lim_{T \to +\infty} p_t(G^T(X), \Gamma^T(X))$$

Example 4.4. Let $\eta, \bar{x}, U$ be as in the Example 2.9. Then

$$i(U) = \sum_{q=0}^{\infty} \dim H^q(S^{m(\bar{x})},*)t^q = t^{m(\bar{x})} \qquad \text{[by 3.9]}$$

since we have
$$H^q(S^k,*) = \begin{cases} 0 & \text{if } q \neq k \\ F & \text{if } q = k \end{cases}$$

Remark 4.4'. By lemma 4.2 (i), $p(N,N_\circ) = p(N/N_\circ,[N_\circ])$; so the generalized Morse index depends only on $h(X)$; thus it is well defined by (1-1) (a) and (b). The above remark implies that the GIM carries less information than the homotopic index. Nevertheless is more useful since it is much easier to deal with. The following theorem illustrates the first properties of the generalized Morse index:

Theorem 4.5. The GIM satisfies the following properties

(i) if $X \in \Sigma$ and for every $x \in X$, there is $t > 0$ such that $x \cdot t \notin X$ then $i(X) = 0$;

(ii) if $X \in \Sigma$ is contractible and positively invariant, then $i(X) = 1$;

(iii) if $X, Y \in \Sigma$ and $X \cap Y = \Phi$ then $i(X \cup Y) = i(X) + i(Y)$;

(iv) if η_i is a semiflow on M_i $(i=1,2)$, then a semiflow $\eta_1 \times \eta_2$ is defined on $M_1 \times M_2$ as follows

$$(\eta_1 \times \eta_2)(t,(x_1,x_2)) = (\eta_1(t,x_1),\eta_2(t,x_2)) ;$$

then if $X_i \in \Sigma(\eta_i)$ $(i=1,2)$, we have that $X_1 \times X_2 \in \Sigma(M_1 \times M_2,\eta_1 \times \eta_2)$ and

$$i(X_1 \times X_2,\eta_1 \times \eta_2) = i(X_1,\eta_1)\cdot i(X_2,\eta_2) .$$

Proof. (i) follows from theorem 1.11; (ii) follows by the fact that

$$H^q(X) = 1 \quad \text{if and only if } q = 0 .$$

(iii) and (iv) follow by lemma 4.3 (ii) and (iii) respectively.//

Next we are going to prove a property of the GIM which is a generalization of the classical Morse inequalities.

Def. 4.6. Take $X_1, X_2 \in \Sigma$ with $\overset{\circ}{X}_1 \cap \overset{\circ}{X}_2 = \Phi$. We say that X_2 is over X_1 if there exists $T > 0$ such that $X_1 \cap G^T(X_1 \cup X_2)$ is positively invariant with respect to $G^T(X_1 \cup X_2)$.

If X_2 is over X_1 or X_1 is over X_2 then we say that X_1 and X_2 are η-connected. Otherwise we say that they are η-disconnected.

Example 4.7. I : If $X_1 \cap X_2 = \Phi$, then X_1 and X_2 are η-disconnected.

II : Let f be a Liapunov function for (M,η) and let c be a constant which is a regular value for f (i.e. $f(x) = c \Longrightarrow f'(x) \neq 0$). We set

$$X_1 = \{x \in M \mid f(x) \leqq c\} ; \quad X_2 = \{x \in M \mid f(x) \geqq c\} .$$

Then $X_1, X_2 \in \Sigma$ and X_2 is over X_1.

Def. 4.8. Let $X \in \Sigma$. A family of sets $\{X_k\}_{k \leqq N}$ is called a Morse decomposition of X if

(i) $X = \bigcup_{k=1}^{N} X_k$

(ii) $X_k \in \Sigma$ for $k=1,\dots,N$

(iii) $\overset{\circ}{X}_k \cap \overset{\circ}{X}_h = \Phi$ for $k \neq h$

(iv) X_{h+1} is over $\bigcup_{k=1}^{h} X_k$ for $h = 1,\ldots,N-1$.

Example 4.9. Let f be a Liapunov function for (M,η) and let $c_1 < c_2 < \ldots$ $\ldots < c_{N-1}$ be a sequence of regular values for f. Let $c_o = -\infty$ and $c_N = +\infty$ and

$$X_k = \{x \in X \mid c_{k-1} \leqq f(x) \leqq c_k\}$$

then $\{X_k\}$ is a Morse decomposition of X.

The next theorem states one for the most important properties of the index (as far as the applications are concerned).

Theorem 4.10. If X_k is a Morse decomposition of X, then there exists $Q \in S$ such that

$$\sum_{h=1}^{N} i(X_k) = i(X) + (1+t)Q(t) \qquad Q \in S .$$

In order to prove Theorem 4.9 some lemmas are necessary.

Lemma 4.10'. Let $X = X_1 \cup X_2$ and suppose that X_2 is over X_1. Then there exist closed spaces $N_o \subset N_1 \subset N_2$ such that (N_2,N_o), (N_2,N_1), (N_1,N_2) are index pairs for X, X_2 and X_1 respectively.

Proof. Take T big enough in order that

(4-4)
(a) $X_1 \cap G^T(X)$ is positively invariant with respect to $G^T(X)$
(b) $(G^T(X),\Gamma^T(X))$ is an index pair for X
(c) $G^T(X_1) \subset \overset{\circ}{X}_1$.

We set

$$N_o = \Gamma^T(X)$$
$$N_1 = (X_1 \cap G^T(X)) \cup \Gamma^T(X)$$
$$N_2 = G^T(X).$$

We want to prove that N_o, N_1, N_2 satisfy the required properties. We nos prove that (N_1, N_o) is an index pair for X. Let us check (i) of Def. 1.1. Since $\overline{N - N_o} = X_1 \cap G^T(X)$

$$G^T(\overline{N - N_o}) = G^T(X_1 \cap G^T(X)) \subset G^T(X_1) \subset \mathring{X}_1 \quad \text{by (4-4) (c).} \tag{4-5}$$

Also by lemma 1.3 (i), (iii) and (iv)

$$G^T(\overline{N - N_o}) \subset G^T(G^T(X)) = G^{2T}(X) \subset \text{int } [G^T(X)] \tag{4-6}$$

Then by (4-5) and (4-6)

$$G^T(\overline{N - N_o}) \subset \text{int}(N - N_o).$$

(iii) of Definition 4.8 holds since $(X_1 \cap G^T(X))$ is positively invariant in $G^T(X)$ by definition and $\Gamma^T(X)$ is positively invariant in $G^T(X)$ by Th. 1.4. Now let us check (iii) of Def. 1.1. If $x \in N_1$ and it leaves N_1 at some times, it has to leave $G^T(X)$ also, since N is positively invariant in $G^T(X)$. Thus there exists t* such that $x \cdot t^* \in \Gamma^T(X)$ is an exit set for $G^T(X)$. Finally since $G^T(X_1) \subset N_1 - N_o$, (iv) of Def. 1.1 holds.

Let us check that (N_2, N_1) is an index pair for X_2.

$$\overline{N_2 - N_1} = \overline{G^T(X) - X_1} = G^T(X) \cap X_2 \ .$$

Then arguing as we have done for $G^T(X) \cap X_1$, it follows that $\overline{N_2 - N_1} \in \Sigma$.

(ii) of Def. 1.1 holds since N_1 is positively invariant in N_2 and (iii) holds since $N_1 \supset \Gamma^T(X)$ and $\Gamma^T(X)$ is an exit set for N_2.

(iv) follows by the fact that $G^T(X_2) \subset \overline{N_2 - N_1}$ .//

<u>Corollary 4.11</u>. If $X = X_2 \cup X_1$ and X_2 is over X_1, then there exists Q such that

$$i(X_1) + i(X_2) = i(X) + (1+t)Q(t) .$$

<u>Proof</u>. By lemma 4.2 (iv) applied to the triple N_o, N_1, N_2 defined in lemma 4.10 we have

$$p(N_2,N_1) + p(N_1,N_o) = p(N_2,N_o) + (1+t)Q(t) .$$

The conclusion follows by lemma 4.10 and the definition of the cohomological index.//

<u>Remark 4.12</u>. It is easy to check that if X_1 and X_2 are η-disconnected, then, for T large enough

$$G^T(X_1 \cup X_2) = G^T(X_1) \cup G^T(X_2) \text{ and } G^T(X_1) \cap G^T(X_2) = \Phi .$$

Then

$$\begin{aligned} i(X) &= i(G^T(X_1 \cup X_2) && \text{by Corollary 1.8} \\ &= i(G^T(X_1)) + i(G^T(X_2)) && \text{by Th. 4.5 (iii)} \\ &= i(X_1) + i(X_2) . \end{aligned}$$

Comparing this result with Corollary 4.11 we deduce that $Q(t) \neq 0$ implies that X_1 and X_2 are η-connected.

Proof of Th. 4.9. We argue by induction. For $N=2$ it is true since it is nothing else but Corollary 2.11.

We can suppose that it is true for N-1; so there exists $Q_1 \in \Sigma$ such that

$$\sum_{k=1}^{N-1} i(X_k) = i(\bigcup_{k=1}^{N-1} X_k) + (1+t)Q_1(t) .$$

Now, since X_N is over $\bigcup_{k=1}^{N-1} X_k$, applying Corollary 4.11 another time, we get

$$i(X_N) + i(\bigcup_{k=1}^{N-1} X_k) = i(X) + (1+t)Q_2(t) \qquad \text{with } Q_2(t) \in \Sigma.$$

Then the conclusion follows with $Q(t) = Q_1(t) + Q_2(t)$ .//

If we have enough compactness we can define the Morse index of an isolated invariant set as follows (cf. also (3-1)).

Def. 4.13. Let S be a (C)-invariant set (cf. Def. 3.1), then we set $i(S) = i(U)$ where $U \in \Sigma$ is a sufficiently small neighbourhood of S.

From the above definition and theorem 4.10 we get

Corollary 4.14. Let X and X_k be as in theorem 4.10.

Moreover suppose that X_k satisfy the property (C) (k=1,...,N) and set $S_k = I(K_k)$. Then we have

$$\sum_{k=1}^{N} i(S_k) = i(X) + (1+t)Q(t) \qquad Q \in S .$$

Observe that in Corollary 1.14 the property (C) for X is not required.

Example 4.15. Let η be a flow as in Example 2.9. Suppose that X and the X_k's satisfy the assumptions of lemma 4.14. Moreover suppose that each X_k contains only one nondegenerate critical point x_k.

Therefore, by the Example 4.4 and Corollary 4.14, we get

$$(4\text{-}7) \qquad \sum_{k=1}^{N} t^{m(x_k)} = i(X) + (1+t)Q(t) \qquad\qquad Q \in S .$$

More in particular, if F(X) = D f(x), then m(x) reduces to the classical Morse index and (4-7) reduces to the classical Morse inequalities.

5. VARIATIONAL SYSTEMS

Let $f \in C^1(M,\mathbb{R})$ and suppose that f' is bounded on bounded sets. If η is a semiflow on M we denote by Df(x) the Dini derivative of f at x i.e.

$$Df(x) = \max \lim_{t \to 0^+} \frac{f(x\ \ t) - f(x)}{t}$$

K(X) will denote the set of critical points of f in X, i.e.

$$K(X) = \{x \in X : f'(x) = 0\} .$$

We need also another notation

$$\Delta_a^b = f^{-1}([a,b]).$$

Def. 5.1. A variational system relative to f is a couple

$\{\eta, \Sigma(\eta)\}$ such that

(i) if $X \in \Sigma$ and $f|_X$ is bounded, then $\forall\, \varepsilon > 0\ \exists\, \delta > 0$ such that

$$Df(x) \leqq -\delta \qquad \forall\, x \in X - N_\varepsilon(K(X)$$

(in particular $Df(x) \leqq 0 \quad \forall\, x \in M$).

(ii) $d(x, \eta(t,x)) \leqq \alpha(|t|)$ where α is a monotone function such that $\alpha(0) = 0$.

(iii) $K(\Delta_a^b)$ is compact.

We recall the condition (c) of Palais and Smale [P.S.] which is essential in constructing variational systems.

Def. 5.2. Let $f \in C^1(M)$. We say that f satisfies (P.S.) if any sequence $\{x_n\}$ such that $f(x_n)$ is bounded and $f'(x_n) \to 0$ is precompact.

If f satisfies (P.S.) and we assume that

$$f \in C^2(M) \qquad \|f'(x)\| \leqq M_1 \tag{5-2}$$

then the equation

$$\dot{x} = -f'(x) \tag{5-3}$$

has a unique solution for every $x \in M$ and $t \in \mathbb{R}$.

If η is the flow relative to (5-3) then it is not difficult to check that $\{\eta, \Sigma\}$ is a variational system for f.

However it is not necessary to assume (5-2) in order to construct a variational system relative to f.

Prop. 5.2 If f satisfies (P.S.) then there exists a variational system $\{\eta,\Sigma\}$ relative to f.

Proof. In [P] Palais has proved that f admits a pseudogradient vector field i.e. a map $F : M \to TM$ such that

(5-4)
(i) the equation $\dot{x} = F(x)$ has a unique solution for every initial point $x \in M$
(ii) $\langle F(x), f'(x)\rangle \geqq \alpha(\| f' \|)$ where α is a strictly monotone function with $\alpha(0) = 0$
(iii) $\| F \|$ is bounded.

Now, if η is the flow relative to the equation (5-4) (i), it is not difficult to prove that $\{\eta, M\}$ is a variational system relative to f.//

We now need a technical lemma:

Lemma 5.3. Let $\{\eta,\Gamma\}$ be a variational system relative to f. Then we have

(i) $T > 0$ such that $G^T(\Delta_a^b)$ is bounded
(ii) let α and β ($a < \alpha \leqq \beta < b$) be two constants such that $K(\Delta_a^\alpha \cup \Delta_\beta^b) = \Phi$. Then $T > 0$ such that $G^T(\Delta_a^b) \subset \Delta_\alpha^\beta$
(iii) if a and b are regular values for f, then $\Delta_a^b \in \Sigma_\circ$ ($\Sigma_\circ$ is defined in § 2)
(iv) if c is the only critical value of $f|_X$ ($X \in \Sigma$, $X \subseteq \Delta_a^b$), then $I(X) = K(X)$ and X satisfies the property (C).

Proof. (i) Take r big enough in order that $K(\Delta_a^b) \subset B_r$ (this is possible by (iii) of Def. 5.1), and set

$$\delta = \inf \{ \| Df(x) \| : x \in \Delta_a^b - B_r \} .$$

By (i) of Def. 5.1, we have $\delta > 0$.

Now take

$$T = \frac{b - a}{\delta}$$

We claim that

(5-5) $$x \in G^T(\Delta_a^b) \implies \bar{t} \in [-T,T] \text{ such that } x \in B_r \cap \Delta_a^b .$$

In fact if $x \cdot [-T,T] \subset \Delta_a^b - B_r$, then $Df(x \cdot t) \leq -\delta$ for every $t \in [-T,T]$.

Therefore

$$b-a \geqq f(x(-T)) - f(x(T)) \geqq \int_{+T}^{-T} Df(x\cdot t)dt \geqq 2T\delta \geqq 2(b-a).$$

The contradiction above implies (5-5).

Now if $x \in \Delta_a^b - B_r$ with $R > r + \alpha(T)$, by (ii) of Def. 5.1 we have that

$$x \cdot [-T,T] \cap B_r = \Phi .$$

By the above formula and (5-5) the conclusione follows.

(ii) It follows easily by (i) and (iii) of Def. 5.1.

(iii) By (i) and (ii) it follows that there exists $\varepsilon, R > 0$ such that

$$G^T(\Delta_{a-\varepsilon}^{b+\varepsilon}) \subset \Delta_{a+\varepsilon}^{b-\varepsilon} \cap B_R .$$

We claim that

$$N_{\varepsilon}(\Delta_{a+\varepsilon}^{b-\varepsilon} \cap B_k) \subset \Delta_a^b$$

Suppose that the above formula does not hold. Then there exist sequences $\phi_n \in \Delta_a^b$ and $z_n \in \Delta_{a+\varepsilon}^{b-\varepsilon} \cap B_R$ such that $d(\phi_n, z_n) \to 0$.

Then, by the mean value theorem we have:

$$\varepsilon < | f(y_n) - f(x_n) | \leqq \| f'(\xi_n) \| \, d(y_n, x_n)$$

The above formula is absurd since we have supposed that f' is bounded on bounded sets.

(iv) Let U be any closed nieghbourhood of K(X). We have to prove that T > 0 such that $G^T(X) \subset U$.

Now let $V \subset U$ be any other closed neighbourhood of K(X) with

(5-6) $d(V, \partial U) > 0$

and set

(5-7) $\delta = \inf\{\| Df(x) \| : x \in X - V\}$.

By (i) of Def. 1.5 we have $\delta > 0$.

Now suppose that $x \in X - U$ and that $x \cdot t \in V$. By (5-6) and (iii) of Def. 5.1, there exists $\tau > 0$ such that

$$|t| > \tau .$$

Then by (5-7)

(5-8) $|f(x) - f(x \cdot t)| \geqq \delta\tau.$

Now we set $\varepsilon = \delta\tau$ and $\delta_1 = \inf\{\| Df(x) \| : x \in (\Delta^{b}_{c+\varepsilon} \cup \Delta^{c-\varepsilon}_{a}) \cap X\}$. Since c is the only critical value of f in X, then $\delta_1 > 0$.

Now we set $T = \frac{b-a}{\delta_1} + \tau$. We claim that $x \notin U \Longrightarrow x \notin G^{T}(X)$. In order to prove the claim above we suppose $f(x) \leqq c$ (if $f(x) \geqq c$ argue in the same way).

By (5-8)we have that $f(x \cdot \tau) \leqq c-\varepsilon$.

Now arguing indirectly we suppose that

$$x \cdot [\tau,T] \subset X .$$

Then we have

$$b-a > f(x\cdot \tau) - f(x\cdot T) \geqq \int_{\tau}^{T} Df(x \quad t)\, dt \geqq (T-\tau)\delta_1 = b-a.$$

The contradiction above implies the conclusion.//

Now set

(5-9) $\mathcal{K}_o = \{K \subset K(M) \mid d(K,K(X)-K) > 0$ and K consists of a finite number of connected components$\}$.

<u>Theorem 5.4</u>. Let $\{M,\eta\}$ be a variational system relative to f. Then

(i) if $K \in \mathcal{K}_o$, then it is a (C)-invariant set; in particular i(K) is well defined.

(ii) if $\{\tilde{\eta},\tilde{\Sigma}\}$ is another variational system, and $K \in \mathcal{K}_o$, then

$$i(K,\eta) = i(K,\tilde{\eta}) .$$

This means that i(K) depends only on f and not on η.

(iii) if $K_1,K_2 \in \mathcal{K}_o$ and $K_1 \cap K_2 = \Phi$, then $i(K_1 \cup K_2) = i(K_1) + i(K_2)$.

(iv) if $X \in \Sigma$, $f|_X$ is bounded below, $K(X) \in \mathcal{K}_o$, then

$$i(K(X)) = i(X) + (1+t)\, Q(t) \qquad Q \in S.$$

Proof. (i) is a trivial consequence of lemma 5.3 (iv).

(ii) To simplify the proof from unessential technicalities we suppose that M is a (may be infinite dimensional) manifold and that η and $\tilde{\eta}$ respectively are the flows relative to the following equations

$$\dot{x} = F(x) \qquad \dot{x} = \tilde{F}(x) .$$

Now let η_λ be the flow relative to the following equation

$$x = (1-\lambda)F(x) + \lambda F(x) \qquad \lambda \in [0,1] .$$

Clearly for every $\lambda \in [0,1]$, $\{M,\eta_\lambda\}$ is a variational system relative to F and K is a (C)-invariant set for η_λ.

Take $\bar{\lambda} \in [0,1]$ and let $U_{\bar{\lambda}}$ be a neighbourhood of K which satisfies the property (C); it exists by (i).

By proposition 3.7, $U_{\bar{\lambda}} \in \Sigma$. Then by the theorem 2.1, $i(U_{\bar{\lambda}},\eta_\lambda)$ is constant for $\lambda \in \Gamma_{\bar{\lambda}}$ where $\Gamma_{\bar{\lambda}}$ is a suitable neighbourhood of $\bar{\lambda}$. This implies that $i(K,\eta_\lambda)$ is constant for $\lambda \in \Gamma_{\bar{\lambda}}$ for every $\bar{\lambda} \in [0,1]$. Thus it follows that

$$i(\eta,K) = i(\eta_0,K) = i(\eta_1,K) = i(\tilde{\eta},K) .$$

(iii) It follows by Theorem 4.5 (iii).

(iv) Since K(X) has a finite number of connected components, $f|_X$ has only a finite number of critical values $c_1,\dots c_N$.

Set

$a_0 = \inf f(X)$

a_ℓ any number in $(c_\ell,c_{\ell+1})$ for $\ell = 1,\dots,N$ and $a_{N+1} = +\infty$.

Now set

$$X_\ell = \Delta_{a_\ell}^{a_\ell+1} \cap X \qquad \text{for } \ell = 0,\dots,N .$$

Then $\{X_\ell\}$ is a Morse decomposition of X (cf. Ex. 4.9). Then by theorem 4.9 we have

$$\sum_{\ell=0}^{N} i(X_\ell) = i(X) + (1+t)\, Q(t) \qquad t \in S . \tag{5-10}$$

By lemma (5.3) (iv), $i(X_\ell) = i(K(X_\ell))$. Using proposition 5.4 (iii), we have

$$i(K(X)) = i(\bigcup_{\ell=0}^{N} K(X_\ell)) = \sum_{\ell=0}^{N} i(K(X_\ell)) = \sum_{\ell=0}^{N} i(X_\ell) .$$

By the above formula and (5-10) the conclusion follows.//

Now we suppose that M is a Hilbert manifold modelled on a space E (i.e. $T_xM \cong E \quad \forall\, x \in M$) and that η is the flow relative to the differential equation

$$\dot{x} = F(x) .$$

Suppose that x is a critical point of F such that $f''(x) : T_xM \to T_xM$ is defined. If

(5-11) $\quad f''(x) : T_xM \to T_xM \qquad$ has a discrete spectrum, we set

(5-12)
$m(x)$ = dimension of the space spanned by the eigenvectors of $f''(x)$ corresponding to negative eigenvalues
$m^*(x) = m(x) + \dim\,[\ker f''(x)]$

We recall that a critical point x is called non-degenerate if $\ker f''(x) = \{0\}$. In this case $m(x) = m^*(x)$.

Theorem 5.5. If x_o is a nondegenerate critical point of f, then $\{x_o\} \in K_o$ and

$$i(x_o) = t^{m(x_o)} .$$

Remark. Observe that in theorem 5.4 we do not assume that f"(x) is defined in a neighbourhood of x_o; it is sufficient that it is defined in x_o. A similar result has been obtained in the contest of the classical Morse theory by Mercuri and Palmieri [Me. P.].

Proof. Since x_o is nondegenerate, it is isolated; thus $\{x_o\} \in K_o$. Now let $\tilde{\eta}$ be the flow relative to the differential equation

$$x = -f''(x_o) \cdot (x-x_o) .$$

If U is a small enough neighbourhood of x_o, then $\{\tilde{\eta},U\}$ is a variational system relative to f.

Then by proposition 5.4 (ii)

$$i(\{x_o\},\eta) = i(x_o,\tilde{\eta}) .$$

But we know by Example 4.4 that $i(x_o,\tilde{\eta}) = t^{m(x_o)}$ .//

The next Corollary follows straightforwardly:

Corollary 5.7. Suppose that $X \in \Sigma$ and that X contains only nondegenerate critical points of f, $x_1,\ldots,x_N$. Then

$$\sum_{k=1}^{N} i(x_k) = i(X) + (1+t)\, Q(t) \qquad Q \in S .$$

Remark 5.8. Notice that in the Corollary above we require only that $f''(x)$ is defined only when x is a critical point of f. This situation occurs quite often when we apply the Morse theory to P.D.E's.

We now need some other notation. If K is a set of critical points of f, then we set

$$(5\text{-}13)\qquad \begin{aligned} m(K) &= \inf_{x\in K} m(x) \\ m^*(K) &= \sup_{x\in K} m^*(x) \end{aligned}$$

The following theorem is quite useful in applications:

Theorem 5.9. Suppose that $U\in\Sigma_o$, that $f|_U$ is bounded and that $f\in C^2(U)$. Then

$$i(U) = \sum_{\ell=m(K)}^{m^*(K)} a_\ell t^\ell \qquad a_\ell \geqq 0$$

where $K = K(U)$.

The proof of theorem 5.9 is based on some results of Marino and Prodi [M. P.] which can be summarized in the following lemma:

Lemma 5.10. Let f, U and K as in theorem 5.9. Then for every $\varepsilon > 0$ there exists a function $g_\varepsilon \in C^2(U)$ such that

(i) $\| f - g_\varepsilon \|_{C^2(U)} \leqq \varepsilon$

(ii) g has only a finite number of critical points in U and they are not degenerate

(iii) all the critical points of g_ε in U are contained in $N_\varepsilon(U)$.

Proof of Th. 5.9. Let $\lambda_1(x) \leqq \lambda_2(x) \leqq \dots \leqq \lambda_k(x) \leqq \dots$ be the eigenvalues of f"(x). They are continuous functions of x in U since f $\in C^2(U)$.

Now let $s = m(K)$ and $r = m^*(K) + 1$.

By the definition of m(K) and m*(K) we have that

(5-14) $\lambda_s(x) < 0 < \lambda_r(x)$ for every $x \in K$.

Now take ε_1 small enough in order that (5-14) holds for every $x \in N_{\varepsilon_1}(K)$. This is possible since the $\lambda_k(x)$ are continuous in x and K is compact. Now let

$$\lambda_1^\varepsilon(x) \leqq \lambda_2^\varepsilon(x) \leqq \dots \leqq \lambda_k^\varepsilon(x) \leqq \dots$$

be the eigenvalue of the operator $g_\varepsilon''(x)$ where g_ε is a function as in lemma 5.10.

Now choose $\varepsilon < \varepsilon_1$ small enough that

(5-15) $\lambda_s^\varepsilon(x) < 0 < \lambda_r^\varepsilon(x)$ $\forall\, x \in N_{\varepsilon_1}(K)$.

Thus we have that all the critical points $x_1,\dots,x_N$ of g_ε are nondegenerate, contained in N_{ε_1} , and by (5-15)

(5-16) $s \leqq m(x_k) \leqq r - 1$ $k = 1,\dots,N$

where $m(x_k)$ is the Morse index of x_k for g_ε.

Now if η_g is the flow relative to the equation

$$\dot{x} = -g'(x)$$

and if ε has been chosen small enough, $U \in \Sigma_0(\eta_g)$ and

(5-17) $i(U,\eta) = i(U,\eta_g)$

by virtue of theorem 2.1.

By Corollary (5.7)

(5-18) $$\sum_{k=1}^{N} i(x_k) = i(U,\eta_g) + (1+t)\,Q(t) \qquad Q \in S$$

where the x_k's are the critical values of g in U.

By theorem 5.5 and 5.13 we have

$$\sum_{k=1}^{N} i(x_k) = \sum_{k=1}^{N} t^{m(x_k)} = \sum_{\ell=s}^{r-1} a_\ell t^\ell .$$

By (5-17), (5-18) and the above formula we have

$$\sum_{\ell=s}^{r-1} a_\ell t^\ell = i(U,\eta) + (1+t)\,Q(t) .$$

From the definition of s and r the conclusion follows.//

<u>Corollary 5.11</u>. If a and b are not critical values for $f' \in C^2(\Delta_a^b)$

$$i(\Delta_a^b) = \sum_{\ell=m(K)}^{m^*(K)} a_\ell t^\ell$$

where $K = K(\Delta_a^b)$.

<u>Proof</u>. Use Theorem 5.9 and Lemma 5.3 (iii) .//

REFERENCES

[B] Benci, V., Some applications of the generalized Morse-Conley index, to appear in "Conferenze del Seminario di Matematica dell'Università di Bari"

[C] Conley, C., Isolated invariant sets and the Morse index, Regional Conference Series in Mathematics, 38, (1976)

[C.Z.] Conley, C., and Zhender, E., Morse type index theory for flows and periodic solutions for Hamiltonian equations, Comm. Pure Appl. Math. 37, (1984), 207-253

[M.P.] Marino, A., and Prodi, G., Metodi perturbativi nella teoria di Morse, Boll. U.M.I., 11, Suppl. Fasc. 3, (1975), 115-132

[Me.P.] Mercuri, F., and Palmieri, G., Morse theory with low differentiability, preprint

[P] Palais, R.S., Ljusternik-Schnirelman theory on Banach manifolds Topology, 5, (1966), 115-132

[P.S.] Palais, R.S., and Smale, S., A generalized Morse Theory, Bull. Am. Math. Soc., 70, (1964), 165-172

[R] Rybakowsky, K.P., On the homotopy index for infinite dimensional semiflows, Trans. Am. Math. Soc., 296, (1982), 351-381

[S] Salamon, D., Connected simple systems and the Conley index of isolated invariant sets, to appear in Trans. Am. Math. Soc.

[Sp] Spanier, E.H., Algebraic topology, Mc Graw-Hill, (1966)

MORSE THEORY AND SYMMETRY

Filomena Pacella

0. INTRODUCTION

One of the results of Morse theory, perhaps the most spectacular, is to obtain the "so called" Morse inequalities which give a relation between the cohomology of a manifold and the critical points of a certain functional defined on that manifold.

In terms of Conley's index this is translated by saying that there is a relation between the cohomology of a space and the "index" of some isolated invariant sets, which form a Morse decomposition for a flow on that space.

The subject of this lecture is to describe what happens to the Morse inequalities when the problem we study (that is the functional or the flow) is subjected to the symmetry of a topological group.

The interesting point, illustrated in the examples of section 3 is that there are (at least) three ways to obtain "Morse relations" and these, generally, give different information.

The subject is divided in three sections. In the first one, after recalling a few definitions about transformation groups, some of the main ideas from Conley's theory are presented.

In section 2 the equivariant Morse theory is described. Finally section 3 deals with very simple examples and ends with an application of the equivariant Morse theory to the N-body problem.

1. PRELIMINARIES: GROUP ACTIONS ON TOPOLOGICAL SPACES AND MORSE INEQUALITIES FOR FLOWS

Let X be a topological space and G a topological group with the multiplicative notation.

Definition 1.1. A left action of G on X is a map:

$$\mu : G \times X \to X \quad , \quad \mu(g,x) = gx$$

satisfying the following properties:

i) $1x = x$, $1 \in G$, $x \in X$

ii) $g_1(g_2x) = (g_1g_2)x$, $g_1, g_2 \in G$, $x \in X$

We speak about a right action if $\mu(g,x) = xg$ and i) and ii) are replaced by:

i)' $x1 = x$, $x \in X$, $1 \in G$

ii)' $(xg_1)g_2 = x(g_1g_2)$, $g_1, g_2 \in G$, $x \in X$.

Given $x \in X$, we denote by $O(x)$ the 'orbit" of x, that is the set of those points in X which can be obtained from x, using the action of the group:

$$O(x) = \{gx, g \in G\}$$

Thus the quotient space X/G is the set of all orbits.

The set $G_x = \{g \in G,\ gx = x\}$, that is the set of the elements in G which leave x fixed, is called the isotropy group of x. If G is a compact topological group, then G_x is a closed subgroup of G.

Definition 1.2. The action of G on X is said to be free if: $g \in G$ and $g \neq 1 \Longrightarrow gx \neq x$, for every $x \in X$.

The definition implies that, when the action is free, $G_x = \{1\}$ for all x in X.

If $x \in X$ and $p : G \to O(x)$ is the map given by $p(g) = gx$, then p is surjective. If the action is free p is also injective which implies that, in this case, every orbit is homeomorphic to G.

Definition 1.3. The action of G on X is said to be effective if:

$$\bigcap_{x\in X} G_x = \{1\}$$

We also define the trivial action of G as the one which leaves everything fixed, that is: $\forall$ x, $G_x = G$.

It is also well known that if G is a compact Lie group acting freely on a manifold X, then X/G is a manifold. However, if the action is not free, or the group is not compact this need not be the case.

Let us recall now some concepts from Conley's theory. For more details and proofs we refer to [5] and [6].

A flow on a topological space X is a map from $X \times R$ onto X, $(x,t) \mapsto x\cdot t$, satisfying the following conditions:

I) $x\cdot 0 = x$, $x\in X$, $0\in R$

II) $(x\cdot s)\cdot t = x\cdot(s+t)$, $x\in X$, $s,t\in R$.

A subset $I\subseteq X$ is said to be invariant if:

$$I = I\cdot R = \{x\cdot t,\ t\in R,\ x\in I\}$$

An invariant set S in X is called an isolated invariant set if it is the maximal invariant set in some compact neighbourhood of itself. Such a neighbourhood is called an isolating neighbourhood for S.

We define the ω-limit sets of $Y\subseteq X$ as:

$$\omega(Y) = \cap\{cl(Y\cdot[t,+\infty))\mid t\geq 0\}$$
$$\omega^*(Y) = \cap\{cl(Y\cdot(-\infty,t])\mid t\leq 0\}$$

Let S be a compact, Hausdorff isolated invariant set in X. A Morse decomposition of S is a finite collection $\{M_\pi\}_{\pi \in P}$ of disjoint, compact, invariant subsets $M_\pi \subseteq S$ which can be ordered $(M_1, M_2, \ldots, M_n)$ in such a way that for every $x \in S \setminus \bigcup_{1 \le j \le n} M_j$ there are indices $i < j$ such that: $\omega(x) \subset M_i$ and $\omega^*(x) \subset M_j$.

The sets M_π, which are also isolated invariant sets as it is easy to demonstrate, will be called Morse sets of S and an ordering of $\{M_\pi\}_{\pi \in P}$ with the above property is called an admissible ordering.

A compact pair (N, N^-) will be called an index pair for the isolated invariant set S if:

a) $cl(N/N^-)$ is an isolating neighbourhood for S

b) $x \in N^-$ and $x \cdot [0,t] \subset N$ imply that $x \cdot [0,t] \subset N^-$

c) if $x \in N$ and $x \cdot R^+ \not\subset N$ then there is a $t \geq 0$ such that $x \cdot [0,t] \subset N$ and $x \cdot t \in N^-$.

The set N^- will be also called the "exit" set of N.

It is possible to prove ([5], [6]) that if (N, N^-) and (N_1, N_1^-) are two index pairs for the isolated invariant set S then the pointed spaces N/N^- and N_1/N_1^- are homotopically equivalent by a homotopy that moves points along orbits of the flow.

Thus to each S there is associated the homotopy class $[N/N^-]$ of the pointed space N/N^- obtained from an index pair. This class will be denoted by h(S) and called the (homotopy) index of S.

We are now ready to state the Morse inequalities.

If $(M_1, \ldots, M_n)$ is an admissible ordering of a Morse decomposition of the isolated invariant set S, then:

$$(1.1) \qquad \sum_{j=1}^{n} P_t(h(M_j)) = P_t(h(S)) + (1+t)\, Q_t$$

where $P_t(h(M_j))$ and $P_t(h(S))$ are the Poincaré series which express the Cech-cohomology (with coefficients in some fixed ring) of any element in the equivalence class $h(M_j)$ or $h(S)$, respectively, and Q_t is a series with nonnegative integer coefficients.

In connection with (1.1) we give the following:

Definition 1.4. A Morse decomposition $(M_1,\ldots,M_n)$ of S is said to be K-perfect if (1.1) holds with $Q_t = 0$, where the cohomology is taken with coefficients in K.

In the particular case of a smooth function f(x) on a compact manifold M of dimension d from (2.1) we obtain the classical Morse inequalities. In fact the equation:

(1.2) $$\dot{x} = -\nabla f(x)$$

defines a gradient flow on M and we can take X = M = S.

Moreover if f has only finitely many critical points, say C = $= \{x_i \mid i = 1,\ldots,n\}$ the collection of the critical points, then C becomes a Morse decomposition of M by ordering its points according to the values of f.

The hypothesis that f has finitely many critical points is verified whenever all the critical points of f are non-degenerate. In the last case we also have:

(1.3) $$P_t(h(x_i)) = t^{d_i}$$

where d_i is the number of negative eigenvalues of the hessian of f in the point x_i, that is the Morse index of x_i.

Regarding the (homotopy) index of S = M, we have:

(1.4) $$P_t(h(S)) = P_t(M) = \sum_{j=0}^{d} \beta_j t^j$$

where $P_t(M)$ is the Poincaré polynomial of M.

Finally, from (1.1), (1.3) and (1.4) we get:

(1.5) $$\sum_{i=0}^{n} t^{d_i} = P_t(M) + (1+t)\, Q_t(f)$$

which are the classical Morse inequalities. The polynomial $\mathcal{M}_t(f) = \sum_{i=0}^{n} t^{d_i}$ is called the Morse polynomial of f.

We end by recalling the definition of non-degenerate critical manifold for a function f and characterizing its Morse index ([1], [3]).

We say that a connected submanifold $T \subseteq M$ is an isolated critical manifold if:

i) each point $p \in T$ is a critical point of f

ii) T is isolated as a critical point set.

From i) and ii) it follows that T is an isolated invariant set in the gradient flow (1.2). Thus T has an (homotopy) index which can be computed as follows when T is non-degenerate. Namely, if the critical manifold T satisfies i) and:

ii)' the hessian of f is non-degenerate in the normal direction to T

then we say that T is a non-degenerate critical manifold. It is obvious that ii)' implies ii), that is each non-degenerate critical manifold is also isolated. Beside ii)' implies that the normal bundle of T, $\nu(T)$, can be decomposed into the direct sum:

$$\nu(T) = \nu^+(T) \oplus \nu^-(T)$$

where $\nu^{+}(T)$ and $\nu^{-}(T)$ are spanned, respectively by the positive and negative eigenvalues of the hessian of f on $\nu(T)$. The fiber dimension, λ_T, of $\nu^{-}(T)$ will be called the (Morse) index of T as a critical manifold of f.

Now, if we want to compute the Poincaré polynomial corresponding to the homotopy-index of T, $P_t(h(T))$ we need to compute the cohomology of N/N^{-}, N being an isolating neighbourhood of T and N^{-} its "exit" set.

But, in the non-degenerate case, the "exit directions" of any isolating neighbourhood are those of $\nu^{-}(T)$, so that:

$$(1.6) \qquad P_t(h(T)) = \sum_i t^i \operatorname{rank} H_c^i(\nu^{-}(T))$$

where H_c^i denotes the compactly supported cohomology[(1)] ([13]).

By the Thom isomorphism:

$$(1.7) \qquad H_c^i(\nu^{-}(T)) = H^{i-\lambda_T}(T, \theta^{-} \otimes K)$$

where K is a ring, θ^{-} is the orientation bundle of $\nu^{-}(T)$ and $H^*(T,\theta^{-}\otimes K)$ is the cohomology with local coefficients. Hence:

$$(1.8) \qquad P_t(h(T)) = t^{\lambda_T} P_t(T, \theta^{-} \otimes K)$$

In particular, when the bundle $\nu^{-}(T)$ is orientable[(2)] $P_t(T,\theta^{-}\otimes K)$

---

(1) If X is a locally compact topological space:
$H_c^i(X) = H^i(\hat{X}) \quad i = 1,2,\ldots$
where $\hat{X}$ is the one point-compactification of X.

(2) We say that a fibration $F \to V \xrightarrow{P} B$

is orientable over a ring K if for any closed path ω in B, with

./.

represents the ordinary cohomology of T with coefficients in the ring K.

2. EQUIVARIANT FLOWS AND EQUIVARIANT MORSE THEORY

Through all this section we assume that X is a Hausdorff topological space and a compact topological group G acts on it.

A subset $Y \subseteq X$ is called G-invariant if:

$$g \in G \quad \text{and} \quad y \in Y \Longrightarrow gy \in Y .$$

We also suppose that an equivariant flow is defined on X, where equivariant means:

$$(2.1) \qquad (gx)\cdot t = g(x\cdot t) \qquad x \in X , \quad g \in G , \quad t \in R$$

This allows us to define a flow on the quotient X/G in the following way:

$$(2.2) \qquad [x]\cdot t = [x\cdot t] \qquad x \in X , \quad t \in R$$

where $[x] = O(x)$ is the orbit of a point $x \in X$, under the action of G.

The flow (2.2) is well-defined because if x' belongs to the orbit of x, then x' = gx, for some g G, and consequently:

$$[x']\cdot t = [x'\cdot t] = [(gx)\cdot t] = [g(x\cdot t)] = [x\cdot t] = [x]\cdot t .$$

./. $\omega(0) = \omega(1) = b \in B$, the induced map:

$$\tau^*_\omega : H^*(F_b;K) \to H^*(F_b;K)$$

is the identity.
In particular, if B is simply connected every fibration over B is orientable, over any K.

If, in particular, we have a gradient flow on X, induced by a function f, then it is equivariant if the function f is G-invariant, that is if: $f(gx) = f(x)$, $x \in M$, $g \in G$.

Let S be a G-invariant isolated invariant set in the G-equivariant flow on X.

The following Proposition is easy to prove (see [12]):

Proposition 2.1. The following statements hold:

i) S/G is an isolated invariant set in X/G

ii) If (N,N^-) is an index pair for S with N and N^- G-invariant then $(N/G,N^-/G)$ is an index pair for S/G

iii) if $(M_1,\ldots,M_n)$ is an admissible ordering of a Morse decomposition of S, given by G-invariant Morse sets, then $(M_1/G,\ldots,M_n/G)$ is an admissible ordering of a Morse decomposition of S/G.

We would like to remark that it is always possible to construct G-invariant index pairs for S ([12]). Besides, if $(M_1,M_2,\ldots,M_n)$ is a Morse decomposition of S, a new Morse decomposition of S, built up by G-invariant Morse sets, is obtained by putting together the set M_i's which are in the same orbit. Actually, if the group is connected each isolated invariant set must be G-invariant.

Let us describe now the equivariant Morse theory for flows ([12])

Since G is a compact topological group there exists an universal G-bundle characterized by having its total space E contractible:

$$G \to E \to E/G = BG \qquad (2.3)$$

The space BG is the so-called classifying space of G. The action of G on E is free and E is unique, up to homotopy.

Since the action of G on E is free, the diagonal action of G on the product $X \times E$ defined by:

$$g(x,e) = (gx,ge) \qquad g \in G\ ,\quad x \in X\ ,\quad e \in E$$

is also free.

The flow can be extended to the product $X \times E$ in the trivial way:

$(x,e)\ t = (x\ t,e) \qquad t \in R.$

Of course this flow is G-equivariant on the product $X \times E$ so that it can be projected to a flow on the quotient space $(X \times E)/G = X_G$.

It is obvious that if S is a G-invariant, invariant set for the flow on X, then $(S \times E)/G = S_G$ is an invariant set for the quotient-flow in S_G.

We would like to obtain some "Morse inequalities" for this quotient flow. To do this, since, usually, E and BG are realized as infinite dimensional manifold, we cannot follow Conley's procedure because we cannot construct compact pairs for any isolated invariant set.

We will use then an approximation method as in [1].

When G is a compact topological group, E and BG can be obtained as limit of finite dimensional compact spaces:

$$E = \lim_{k \to \infty} E_k \qquad\qquad BG = \lim_{k \to \infty} B_k G$$

related to the bundles:

$$\begin{array}{c} G \\ \downarrow \\ E_k \\ \downarrow \\ E_k/G = B_k G \end{array}$$

Here the action of G on E_k is free which implies that the diagonal action in the product $X \times E_k$ is also free.

As before we extend the flow to the product $X \times E_k$ in the trivial way and project it to the quotient space $(X \times E_k)/G$.

Let S be a G-invariant isolated invariant set in X and $(M_1,\ldots,M_n)$ a Morse decomposition of S built up by G-invariant sets.

As a consequence of Proposition 2.1 we have that $(S \times E_k)/G$ is an isolated invariant set for the flow on $(X \times E_k)/G$ and $((M_1 \times E_k)/G,\ldots,(M_n \times E_k)/G)$ is an admissible ordering of a Morse decomposition of $(S \times E_k)/G$.

Then for each $k \in N$, we can write the following Morse inequalities

$$(2.4) \qquad \sum_{j=1}^{n} P_t(h_k(M_j)) = P_t(h_k(S)) + (1+t)\, Q_t^k$$

where $h_k(S)$ and $h_k(M_j)$ are, respectively, the homotopy-index of $(S \times E_k)/G$ and $(M_j \times E_k)/G$.

In particular $P_t(h_k(S))$ (resp. $P_t(h_k(M_j))$ can be computed by any index pair $((N \times E_k)/G, (N^- \times E_k)/G)$ of $(S \times E_k)/G$ (resp. of $(M_j \times E_k)/G$) if (N,N^-) is a G-invariant index pair for S (resp. M_j).

Now, if we take the limit in (2.4), $k \to \infty$, using the stabilization property of the cohomology for the classifying spaces of compact topological groups ([7])[(3)] we obtain:

$$(2.5) \qquad \sum_{j=1}^{n} P_t^G(h(M_j)) = P_t^G(h(S)) + (1+t)\, Q_t^G .$$

In (2.5) $P_t^G(h(S))$ (resp. $P_t^G(h(M_j))$) represents the cohomology of the pair $((N \times E)/G, (N^- \times E)/G)$, if (N,N^-) is a G-invariant index pair for S (resp. M_j), that is equivariant cohomology of the pair (N,N^-).[(4)]

---

(3) If $E = \lim_{k\to\infty} E_k$ and $BG = \lim_{k\to\infty} E_k/G$ then: for each $i \in N$ there exists $m(i) \in N$ such that:
$k \geq m(i) \Rightarrow H^i(E) = H^i(E_k)$ and $H^i(BG) = H^i(E_k/G)$

(4) The equivariant cohomology of a space Y with respect to the group G is the cohomology of $(Y \times E)/G = Y_G$

Denoting by $h_G(S)$ (resp. $h_G(M_j)$) the homotopy type of the pair $((N \times E)/G, (N^- \times E)/G)$, (2.5) could be written as:

$$(2.6) \qquad \sum_{j=1}^{n} P_t(h_G(M_j)) = P_t(h_G(S)) + (1+t)\, Q_t^G .$$

It is possible to prove ([12]) that, if the action of G is free on a G-invariant isolated invariant set I, then $P_t(h_G(I))$ is equal to $P_t(h(I/G)$ where $h(I/G)$ is the (homotopy) index of I/G in the quotient flow on X/G.

In the particular case of a gradient flow, induced by a non-degenerate G-invariant smooth function f on G-invariant compact manifold M, (2.5) gives the equivariant Morse inequalities described in [1] and [3].

Since a non-degenerate G-invariant function has only a finite number of critical orbits, which are non-degenerate critical manifolds we can take as a Morse decomposition of f the one given by these critical orbits, ordering them according to the values of f.

From (2.5) we get:

$$(2.7) \qquad \sum_{T} P_t^G(h(T)) = P_t^G(M) + (1+t)\, Q_t^G$$

where the summation is extended to all the critical orbits of f. But, for non-degenerate critical manifolds the (homotopy) index can be computed by formula (1.8). So, using, for example, the above approximation method, we have:

$$(2.8) \qquad \sum_{T} t^{\lambda_T} P_t^G(T, \theta^- \otimes K) = P_t^G(M) + (1+t)\, Q_t^G$$

where $P_t^G(M)$ (resp. $P_t^G(T)$) represents the equivariant cohomology of M (resp. T) and λ_T is the number of negative eigenvalues of the hessian of f in the direction normal to T.

Moreover each T, being a single orbit, is homeomorphic to G/H, where H is the isotropy group of each point of T.

Hence we have:

$$(T \times E)/G \simeq ((G/H) \times E)/G \simeq E/H \simeq BH$$

The last homotopy-equivalence between E/H and BH (which is the classifying space of H) holds because if E is the total space for a universal G-bundle, E is also the total space for a universal H-bundle.

Then (2.8) becomes:

$$(2.9) \qquad \sum_T t^{\lambda T} P_t(BH, \theta^- \otimes K) = P_t^G(M) + (1+t)\, Q_t^G$$

Finally, if H is connected, we do not need to use local coefficients because, in this case, BH is simply connected and every bundle over a simple connected base is orientable.

3. EXAMPLES

Let S be a G-invariant isolated invariant set for a flow on a Hausdorff topological space X, equivalent under the action of a compact topological group G.

From the two previous sections we deduce that there are, at least three ways of obtaining "Morse relations" for S.

A first one could be to study the "quotient flow" on S/G and apply there Conley's theory.

A second approach would be to look at the isolated invariant set S but considering Morse decomposition whose Morse sets contain the complete G-orbit of any point in the set (if the action of G is not free these orbits may be topologically different).

A third approach is given by the "equivariant Morse theory".

When the action of G on S is free there is not much difference between these three methods; in particular the first and the third one give exactly the same answer because, in this case, the equivariant cohomology coincides with the ordinary cohomology.

When the action is not free, then in general each method furnishes different information, that, is, the Morse inequalities provide different consistency conditions.

To understand this difference it is enough to think about the difference between X, X/G, $(X \times E)/G$ at the cohomological level. It may happen that X has a trivial cohomology while X/G has a rich cohomology and vice versa. For instance, if S^∞ is the sphere in a Hilbert space and S^1 acts on it with the Hopf action, then $P_t(S^\infty) = 1$ because S^∞ is contractible while $P_t(S^\infty/S^1) = P_t(CP^\infty) = 1+t^2+\ldots+t^{2n}+\ldots = 1/(1-t^2)$.

Moreover if we have a gradient-flow on a compact G-invariant manifold M, induced by a G-invariant non-degenerate function f and if the action on M is not free then the classical Morse theory does not apply because M/G is not, in general, a manifold. The theory of Conley does apply, but gives different information from the equivariant theory.

We will support what we have claimed with some examples.

Before doing that let us remark that, actually, if S is an isolated invariant set in a local flow, two (homotopy) indexes are defined, according to the two directions of the time.

The first one, in the forward direction is the one already defined. The second, in backward time, can be defined "reversing" the flow with respect to the time. This means that we consider an index pair (N,N^+), where N^+, the "entrance" set, is defined by the properties dual with respect to those which define N^-.

In the gradient-flow case this is realized by considering -f instead of f. Consequently, considering a Morse decomposition of S, we have two different kinds of Morse inequalities, according to the two different indexes of the Morse sets. This, in general, gives more information. For example, suppose the isolated invariant set S is the total space. Then the indexes in the two different directions are the same. Now, if the Poincaré polynomial $P_t(h(S))$ is not symmetric,[(5)] different information comes from the two sets of Morse relations. Of course if M is a compact manifold (without boundary) then, from the Poincaré duality Theorem ([13]), its Poincaré polynomial is symmetric, but, since the Morse theory applies also to manifolds with boundary (or general compact metric spaces) the consideration of the index in both directions can be really useful. This happens, in particular, when we have a quotient space M/G, where M is a manifold and G does not act freely on M, as we will see in the next examples.

Example 3.1. Let S^1 be the unit circle and $S^2 = \{(x,y,z) \in R^3, x^2+y^2+z^2 = 1\}$ the unit sphere in R^3. Identifying the point $(x,y,z) \in R^3$ with the point $(x+iy,z) \in C \times R$, we consider the action of S^1 on S^2 given by $\zeta(x+iy,z) = (\zeta(x+iy,z)$. This is a rotation about the z-axis. This action is not free because the points $P_1 = (0,0,-1)$, $P_2 = (0,0,+1)$ are fixed. For the other points of S^2, instead the isotropy group is $\{1\}$.

Let f be the function:

$$f(x,y,z) = z \qquad \text{on } S^2.$$

The only two critical points of f are the txo fixed points P_1 and P_2 (resp. min. and max. of f).

---

(5) We say that the polynomial : $a_o + a_1 t + \ldots + a_n t^n$ is symmetric if $a_i = a_{n-i}$.

Let us examine the three different approaches:

a) First of all we consider the quotient space S^2/S^1 which is homeomorphic to the interval $[-1,1]$ on the z-axis. This is a contractible set and hence has a trivial cohomology: $P_t(S^2/S^1) = 1$.

So it seems that from this cohomology we can just guess the presence of one critical point, that is the minimum of f.

But if we reverse the flow with respect to the time direction, that is, if we consider -f instead of f we discover another critical point. In fact, since $P_t(S^2/S^1)$ does not change also -f has to have a minimum that cannot be the same as the one of f.

On the other hand we know that there are two critical points and that one is an attractor and one is a repeller for the gradient flow on S^2/S^1. The (homotopy) indexes are:

$$h(P_1) = \overline{1} \quad \text{and} \quad h(P_2) = \overline{0}$$

that is $h(P_1)$ corresponds to the homotopy type of the pointed 0-sphere and $h(P_2)$ corresponds to the homotopy type of the pointed one-point space (see [5]).

Hence, considering the Morse decomposition (P_2,P_1) we have:

$$P_t(h(P_2)) + P_t(h(P_1)) = 1 + 0 = P_t(S^2/S^1) = 1 .$$

Consequently (P_2,P_1) is a perfect Morse decomposition of S^2/S^1.

b) Here we consider the function directly on S^2 and look at the critical orbits. These consist of P_1 and P_2, since these two points are fixed under the action of S^1.

The Morse indexes, as number of negative eigenvalues, of P_1 and P_2 are 0 and 2, respectively. The cohomology of S^2 is: $P_t(S^2) = 1+t^2$. Then we have:

$$\mathcal{M}_t(f) = 1+t^2 = P_t(S^2) = 1+t^2$$

that is f is still a perfect function.

c) Finally we use the equivariant approach. Looking at the fibration:

$$(3.1) \qquad \begin{array}{c} S \\ \downarrow \\ (S^2 \times E)/S^1 \\ \downarrow \\ BS^1 \end{array}$$

E being the total space of an universal bundle of S^1, we have:

$$(3.2) \qquad P_t^{S^1}(S^2) = P_t(S^2) \cdot P_t(BS^1) = (1+t^2)/(1-t^2).$$

We have obtained the product formula (3.2) from the spectral sequence associated to (3.1) observing that the classifying space of S^1 is the infinite-dimensional complex projective space whose cohomology is $1+t^2 +...+ t^{2n} +...$

Since every critical manifold of f consists of a single orbit (namely P_1 or P_2) and the isotropy group is S^1, we can apply (2.9), with $BH = BS^1$.

Then we obtain:

$$\mathcal{M}_t^{S^1}(f) = 1/(1-t^2) + t^2/(1-t^2) = P_t^{S^1}(S^2) = (1+t^2)/(1-t^2).$$

Hence f is equivariantly perfect.

Let us observe that in this case, as in the previous one, reversing the flow nothing changes.

Example 3.2. We consider the same action as in the previous example and the function: $f(x,y,z) = z^2$ on S^2.

This function has a minimum corresponding to the circle orbit at $z = 0$ and two maxima corresponding to the points with $z = \pm 1$.

We have:

a) In the quotient space S^2/S^1 the point P_o with $z = 0$ is an attractor $P_t(h(P_o)) = 1$, P_1 and P_2 are both repellers, $P_t(h(P_j)) = 0$, $j = 1,2$.

Hence, considering the Morse decomposition (P_1,P_2,P_o) we obtain:

$$P_t(h(P_j)) = 1 = P_t(S^2/S^1)$$

that is the Morse decomposition is perfect.

If we reverse the flow, the P_o becomes a repeller and P_1 , P_2, both attractors. The associated Morse decomposition is (P_o,P_1,P_2) and we have: $P_t(h(P_o)) = t$, $P_t(h(P_j)) = 1$, $j = 1,2$.

Thus the Morse inequalities are:

$$\sum_{j=0}^{2} P_t(h(P_j)) = t + 2 = P_t(S^2/S^1) + 1 + t.$$

Therefore (P_o,P_1,P_2) is not a perfect Morse decomposition.

b) Considering f on S^2 we have a critical orbit homeomorphic to S^1 corresponding to the minimum whose contribution in the Morse inequalities, according to (1.8) is: $t^o P_t(S^1) = 1+t$.

The other two critical orbits are the points P_1 and P_2 whose Morse index is 2 (non-degenerate maxima).

So we have:

$$M_t(f) = (1+t) + 2t^2 = P_t(S^2) + (1+t)Q_t = 1+t^2 + (1+t)t \ .$$

This means that, using this approach, f is not perfect and $Q_t = t$.

Reversing the flow we have:

$$\dot{M}_t(f) = 2 + t(1+t) = P_t(S^2) + 1 + t$$

that is f is still not perfect and $Q_t = 1$.

c) Using the equivariant theory and considering that the isotropy group of each point of the circle orbit is $\{1\}$, (1 is the unity element in the group S^1), we get:

$$M_t^{S^1}(f) = 1 + (2t^2)/(1-t^2) = (1+t^2)/(1-t^2) = P_t(S^2)$$

hence f is perfect.

If we reverse the flow, then we have:

$$M_t^{S^1}(f) = 2/(1-t^2) + t = (1+t^2)/(1-t^2) + (1-t)$$

therefore f is not perfect and $Q_t = 1$.

Example 3.3. Let $q_1,\ldots q_N \in R^3$ denote the position of N-bodies with masses $m_1,\ldots,m_N$ respectively. The motion of these N-bodies is described by Newton's equations:

$$(3.3) \qquad M\ddot{q} = -\nabla V(q)$$

where $q = (q_1,\ldots,q_N)$, $V(q) = -\sum_{i<j} \frac{m_i m_j}{|q_i - q_j|}$ is the potential energy and

$$M = \begin{pmatrix} m_1 I_3 & & \\ & \ddots & \\ & & m_N I_3 \end{pmatrix}.$$

From the equations it follows that $\sum_i m_i q_i = 0$ which is equivalent to fix the center of mass in the origin of R^3.

The configuration space of these N-bodies is the space $X = \{q \in R^{3N}, \sum_i m_i q_i = 0\}$ and $\Delta = \{q \text{ s.t. } \exists\, i \neq j,\ q_i = q_j\}$ is the collision set.

<u>Definition 3.1. A point</u> $\bar{q} = (\bar{q}_1,\ldots,\bar{q}_N) \in X\setminus\Delta$ <u>is called a central configuration if there exists a solution of (3.3) in the form</u> $\phi(t)\bar{q}$, <u>with</u> $\phi(t)$ <u>scalar function</u>.

It is possible to prove that this definition is equivalent to say that $\bar{q}$ is a critical point of the potential V restricted to the manifold $\Sigma = E\,\Delta$, where $E = \{q|(q,Mq) = 1\}$ ([10]).
Then we could use Morse theory in order to investigate the number of central configurations.

It is easy to see that the manifold Σ is invariant with respect to the diagonal action of the orthogonal group O(3). Also the potential V(q) is O(3)-invariant.

This action is not free. In fact every collinear configuration has O(2) as isotropy group, while every coplanar configuration has Z_2 as isotropy group.

The presence of different (not finite) isotropy groups brings in some commplications. The first one is that we do not know the cohomology of the quotient space $\Sigma/O(3)$.

In [10] the central configurations are studied using the equivariant Morse theory. In order not to use local coefficients only the action of SO(3) has been considered because that avoids the presence of disconnected isotropy groups. In fact in this case, every collinear configuration has $SO(2) = S^1$ as isotropy group, while the action is free on the other types of configurations.

The first thing to compute is the SO(3)-equivariant cohomology of Σ. This can be done with some tools of algebraic topology, knowing the cohomology of the classifying space BSO(3), with rational coefficients, and the cohomology of Σ ([10]).

Then we have:

$$(3.4) \qquad P_t^{SO(3)}(\Sigma) = \frac{(1+t^2)(1+2t^2)\ldots(1+(N-1)t^2)}{1-t^4}$$

As pointed in Example 3.1 the classifying space of S^1, which represents the only nontrivial isotropy group, is the infinite dimensional complex projective space.

So, using also a precise estimate of the index of the collinear central configurations ([10]), by (2.9) we get:

$$(3.5) \qquad \sum_{\lambda=0}^{2N-5} (\alpha_\lambda+\beta_\lambda)t^\lambda + (N!/2)t^{2N-4}/(1-t^2) = P_t^{SO(3)}(\Sigma) + (1+t)Q_t(V)$$

where α_λ and β_λ are the number of critical orbits corresponding to the spatial and coplanar configurations (respectively) and $(N!/2)t^{2N-4}/(1-t^2)$ is that part of the Morse polynomial of V relative to the collinear central configurations.

From (3.4) and (2.5) we deduce immediately an estimate of the number of spatial and planar central configurations (see [10] for further details).

4. REFERENCES

[1] Atiyah, M.F., Bott, R.:"The Yang-Mills equations over Riemann surfaces", Phil. Trans. R. Soc. London, A 308, (1982) 523-615

[2] Benci, V., Pacella, F.:" Morse theory for symmetric functionals on the sphere and an application to a bifurcation problem", Nonlinear Analysis T.M.A. vol. 9, 8, (1985) 763-773

[3] Bott, R.: "Lectures on Morse theory, old and new",Bull. Am. Math. Soc. vol. 7 N. 2, (1982) 331-358

[4] Bredon, E.G.: "Introduction to compact transformation groups", Academic Press, New York (1972)

[5] Conley, C.C.: "Isolated invariant sets and the Morse index", CBMS Regional Conf. Series in Math. 38 A.M.S. Providence R.I. (1978)

[6] Conley, C.C., Zehnder, E.: "Morse type index theory for flow and periodic solutions for hamiltonian equations", Comm. Pure and Appl. Math. (1984)

[7] Hofmann, K.H., Mostert, P.S.: "Cohomology theories for compact abelian groups", Springer Verlag, New York (1973)

[8] Husemoller, D.: "Fibre bundles" Springer Verlag, New York (1966)

[9] Milnor, J.: "Morse theory", Ann. Math. Studies 51, Princeton Univ. Press, Princeton (1963)

[10] Pacella, F.: "Central configurations of the N-body problems via the equivariant Morse theory", Arch. Rat. Mech. Anal. (to appear)

[11] Pacella, F.: "Morse theory for flows in the presence of a symmetry group", M.R.C. Rep. N. 2717, July 1984

[12] Pacella, F.: "Equivariant Morse theory for flows and an application to the N-body problem",Trans. Amer. Math. Soc. (to appear)

[13] Spanier, E.H.: "Algebraic topology", McGraw-Hill, New York (1966)

[14] Wassermann, A.G.: "Equivariant differential topology", Topology 8 (1969) 127-150

DYNAMICAL SYSTEMS WITH SINGULAR POTENTIALS

Antonio Ambrosetti(*)
Scuola Normale Superiore, Pisa, Italy

In this talk we will discuss some results dealing with the existence of solutions with prescribed period $T > 0$ to conservative second order systems:

$$(1) \qquad -\ddot{p} = V'(p) \qquad (V'(p) = \text{grad } V(p))$$

when the potential V has singularities on the boundary $\partial\Omega$ of an open set $\Omega \subset \mathbb{R}^n$, in the sense that $V \in C^2(\Omega, \mathbb{R})$ and $V(x) \to -\infty$ (or $+\infty$) as $x \to \partial\Omega$.

The interest for this kind of problems comes, among others, from the fact that singular potentials arise frequently in Classical (or Celestial) Mechanics, for example.

<u>Notations</u>: For $x,y \in \mathbb{R}^n$, $x \cdot y$ denotes the Euclidean scalar product in $\mathbb{R}^n$ and $|x|^2 = x \cdot x$. We let $S^{n-1} = \{x \in \mathbb{R}^n : |x| = 1\}$.

$\int p = \int_0^T p(t)dt$

(*) Supported by Ministero P.I., Gruppo Naz. (40%) "Calcolo delle Variazioni"

$$\| p \|^2 = \frac{1}{T} \int_0^T |p|^2$$

$$\| p \|_1^2 = \| p \|^2 + \| \dot{p} \|^2$$

The strong (weak) convergence is denoted by $\to$ (resp. $\rightharpoonup$).

A. We begin discussing a results by W.B. Gordon [1]. For the sake of simplicity, we will suppose:

(2) $\quad V \in C^2(\Omega, \mathbb{R}) \quad , \quad \Omega \subset \mathbb{R}^n$ open;

(3) $\quad V(x) \leqq 0 \qquad \forall\, x \in \Omega$.

In general, the T-periodic solutions of (1) are the critical points of the functional

(4) $\quad F(p) = \frac{1}{2} \| \dot{p} \|^2 - \int V(p)$

on

$$\Lambda : = \{ p \in H^{1,2}(S^1; \mathbb{R}^n) : p(t) \in \Omega \quad \forall\, t \}$$

1. Remarks. a) Here a control on the behaviour of F on the boundary $\partial\Lambda$ of Λ is needed. Let us notice that F could be finite on $\partial\Lambda$. For example, letting $\Omega = \mathbb{R}^n \setminus \{0\}$ and $V(x) = - |x|^{-a}$, with $0 < a < 2$, and taking $p \in H^{1,2}(S^1, \mathbb{R}^n)$ with $p(t) = |t|^k x^*$, $\frac{1}{2} < k < \frac{1}{a}$, $x^* \in \Omega$, for $|t|$ small, one has ($p \in \partial\Lambda$ and) $F(p) < + \infty$.

b) In general F is neither coercive on Λ , nor the condition (c) of Palais-Smale is satisfied. For example, taking Ω and V as in point a), then for every sequence $x_j \in \mathbb{R}^n$, with $|x_j| \to +\infty$, one has $F(x_j) \to 0$ and $F'(x_j) \to 0$.

Overcoming the difficulties sketched in Remarks 1. W.B. Gordon has shown:

2. Theorem. Suppose (2) - (3) hold. Further, we assume:

(SF) $\exists$ $U \in C^2(\Omega, \mathbb{R})$, $U(x) \to -\infty$ as $x \to \partial\Omega$, such that $V(x) \leq -|U'(x)|^2$;

(Ω) Ω contains a regular cycle p^* verifying the following property: $\forall\ c > 0\ \exists\ K_c \subset \mathbb{R}^n$, compact, which contains every cycle q which is homotopic to p^* in Ω, whose length is $\leq c$.

Then $\forall T > 0$ (1) has one T-periodic solution.

Let us point out that (Ω) implies that the singularities of V must be rather "complicated". For example, if $\Omega = \mathbb{R}^n \setminus \{0\}$ then (Ω) holds if and only if $n = 2$.

Coming back to Theorem 2, the proof is based, roughly, on the following arguments: first of all (SF) (= "Strong Force") permits to show that $F(p) \to +\infty$ as $p \to \bar{p} \in \partial\Lambda$. This fact, jointly with (Ω), allows to prove: (i) F is coercive of the component Λ^* of Λ such that $p^* \in \Lambda^*$; (ii) $q \in \Lambda^*$ such that $F(q) = \min \{F(p): p \in \Lambda^*\}$.

An extension to potentials which are time-dependent has been given recently by Capozzi, Greco and Salvatore [2].

B. If one pursues an existence result for (1) without making an assumption like (Ω), a different argument is needed: in fact (cfr. Remark 1b) the condition (C) does not hold any more.

We will suppose (again for simplicity) that (2) - (3) hold. We shall assume also (SF).

On the behaviour of V at infinity, we make the following hypothesis:

(4) $V(x)$ and $V'(x) \to 0$ as $|x| \to +\infty$ and $\exists\, r > 0$:
$x \cdot V'(x) > 0 \quad \forall\, |x| > r.$

From (4) one deduces:

3. Lemma. Let (2) - (3) - (4) and (SF) hold. Then:

(i) $\forall\, \varepsilon > 0$ condition (C) is verified on $A_\varepsilon := \{ p \in \Lambda : F(p) \geqq \varepsilon \}$;

(ii) if $\varepsilon > 0$ is small enough, then $H_k(A_\varepsilon) = H_k(S^{n-1})$.

Sketch of the proof. Let $p_j \in \Lambda$ be such that

(5) $\varepsilon \le F(p_j) \le \text{const.}$

(6) $F'(p_j) \to 0$

Letting $\xi_j = \int p_j$ and $w_j = p_j - \xi_j$, (5) gives:

(7) $\| \dot{w}_j \| \leqq \text{const.}$

If, in addition to (7), $|\xi_j| \leqq$ const, then it follows readily that $p_j \to p^*$ ($p^* \in \Lambda$ by (SF)) and (i) would be true.

To show that $|\xi_j| \leqq$ const, we argue by contradiction. If $|\xi_j| \to +\infty$, (7) implies $|p_j(t)| \to \infty$ uniformly in t and thus, by (4):

(8) $V(p_j(t)) \to 0$ and $V'(p_j(t)) \to 0$ uniformly in t

From

$$\| \dot{W}_j \|^2 = \int V'(p_j).W_j + (F'(p_j),W_j)_{H^1}$$
$$\leqq \| \dot{W}_j \|_{L^\infty} \| V'(p_j) \|_{L^1} + \| \dot{W}_j \| \, \| F'(p_j) \|_1$$

and (6), (7), (8), one finds:

(9) $\qquad \| \dot{W}_j \|_j \to 0$

Lastly (8) and (9) imply that $F(p_j) = \frac{1}{2} \| \dot{W}_j \|^2 - \int V(p_j) \to 0$, in in contradiction with (5). This shows that $|\xi_j| \leq$ const and hence (i) holds.

To prove (ii) one defines a deformation of A_ε on

$$\{\xi \in \mathbb{R}^n : |\xi| \geqq r\} \qquad r > 0 \text{ large}$$

Roughly, one first deforms A_ε on $A_{\varepsilon'}$ $(\varepsilon' < \varepsilon)$ using the steepest descent flow; then each $p \in A_{\varepsilon'}$ is mapped on $\xi = \int p$ through the projection

$$\Pi(s,p) = sw + \xi, \quad 0 \leq s \leq 1 \quad , \quad w = p - \xi$$

One shows that for ε small enough $\Pi(s,p)$ $A_\varepsilon \in \forall s \in [0,1]$ whenever $p \in A_{\varepsilon'}$, with $\varepsilon' \ll \varepsilon$, and (ii) follows.

One way to use Lemma 3 is to suppose that $\mathbb{R}^n - \Omega$ is compact. In such a case $H_*(\Lambda)$ is infinite and one can show:

4. <u>Theorem</u>. Suppose that (2), (3), (4) and (SF) hold and let $\mathbb{R}^n - \Omega$ be compact. Then $\forall\, T > 0$, (1) has infinitely many T-periodic solutions.

An additional argument permits to prove that, under rather general conditions, such solutions are non constant and geometrically distinct.

Theorem 4 is a particular case of more general results contained in a joint paper with V. Coti Zelati [3].

We refer to such a paper for more details as well as for a discussion of an abstract critical point theory which applies to study (1) and all classes of dynamical systems (singular and not).

For other results on problems with singular potentials we refer: (i) V. Coti Zelati [4] where $V(x)$ is of the type $|x|^{-2} - |x|^{-1}$ ("effective-like" potentials); and (ii) C. Greco [5] who deals with potentials which can tend to $+\infty$ as $|x| \to +\infty$.

C. In all the above results, the (SF) condition is always assumed. If (SF) does not hold the situation becomes more complicated.

In fact, not only F can have finite value on $\partial\Lambda$ (cf. Remark 1a), but F could even possess critical points on $\partial\Lambda$. For example, if $V(x) = -|x|^{-1}$, one finds the so-called "legs" (cfr. [6]) corresponding to segments with one vertex in $0 \in \mathbb{R}^n$.

5. Remark. From the conservation of the energy:

$$\frac{1}{2}|\dot{p}(t)|^2 + V(p(t)) = E \tag{10}$$

it follows that there can exist solutions of (1) which touch $\partial\Omega$ only in the case in which $V(x) \to -\infty$ as $x \to \partial\Omega$.

However, it can be somewhat interesting to show that such singular solutions can be avoided (even if $V(x) \to -\infty$ as $x \to \partial\Omega$) in some situations by means of a technique used in a preceding paper [7] to study systems like (1) in a potential well.

Let $\Omega \subset \mathbb{R}^n$ be (open) bounded and convex and suppose $V \in C^2(\Omega, \mathbb{R})$ be such that $V(x) \to -\infty$ as $x \to \partial\Omega$. Moreover we assume that V is strictly concave (this is for simplicity only: a much weaker condition is actually needed [9]). Letting G the Legendre transform of -V, one has that $G \in C^2(\mathbb{R}^n, \mathbb{R})$, is convex and $G(y) \gtrsim |y|$ for $|y|$ large. Set

$$X = \{ u \in L^1(0,T:\mathbb{R}^n) : \int u = 0 \}$$

We define $L : x \to x$ by $Lu = z$ iff $-\ddot{z} = u$.

Consider the functional $\Phi \in C^1(x, \mathbb{R})$,

$$\Phi(u) = \int [G(u) + \frac{1}{2} u.Lu]$$

Notice that (G makes sense in all $\mathbb{R}^n$, hence) Φ is defined on all space X, in spite of the fact that V was defined on the open set Ω only.

If $\Phi'(u) = 0$ then $\exists \xi \in \mathbb{R}^n$ such that $G'(u) + Lu = \xi$. Setting $z = Lu - \xi$ $(= -G'(u))$ one has that z is a T-periodic solution of (1) (Dual Action Principle, introduced in [8] when $\Omega = \mathbb{R}^n$).

Now, from $z = Lu - \xi$ it follows that $z \in W^{2,1}$, in particular $\dot{z}$ is A.C. From $z = G'(u)$ $(u \in L^1)$ it follows that $z(t) \in \Omega$ for almost all t. If I is a compact interval with the property that $z(t) \in \Omega$ $\forall t \in I$, (10) and the fact that $|\dot{z}(t)|^2$ is bounded, imply that $V(z(t)) \leq c$, where c is a constant independent on I. As consequence one has that $z(t) \in \Omega$ $\forall t$.

In this way it is possible to find periodic solutions of (1) without assuming any "strong-force" condition. For more details, we refer to [9], where non-concave and/or time-dependent potentials are also studied.

It would be interesting to extend in a suitable way the preceding approach to study the case of the Newtonian potentials.

References

1. Gordon, W.B.: Conservative Dynamical Systems involving strong forces. Trans. Amer. Math. Soc. 204 (1975)
2. Capozzi, A., Greco, C.,and Salvatore, A.: Lagrangian Systems in presence of singularities. Preprint
3. Ambrosetti, A. and Coti Zelati, V.: Critical points with lack of compactness and singular dynamical systems. Preprint
4. Coti Zelati, V.: Conservative systems with effective-like potentials. Nonlinear Analysis TMA, to appear
5. Greco, C.: Periodic solutions of some nonlinear ODE with singular nonlinear part. Preprint
6. Gordon, W.B.: A minimizing property of Keplerian orbits. Amer. Jour. Math. 99 (1977)
7. Ambrosetti, A. and Coti Zelati, V.: Solutions with minimal period for Hamiltonian systems in a potential well. Annales de l'IHP, Analyse Nonlineaire, in corso di stampa
8. Clarke, F. and Ekeland, I.: Hamiltonian trajectories having prescribed minimal period. Comm. Pure Appl. Math. 33 (1980)
9. Coti Zelati, V.: Remarks on dynamical systems with weak forces, preprint.

SUBHARMONIC SOLUTIONS OF PRESCRIBED MINIMAL PERIOD FOR NON AUTONOMOUS DIFFERENTIAL EQUATIONS

Vieri Benci - Donato Fortunato

Istituto di Matematiche Applicate - Università di Pisa
Dipartimento di Matematica - Università di Bari
ITALY

1. STATEMENT OF THE RESULTS

Let $V = V(x,t)$ $(x \in \mathbb{R}^N,\ t \in \mathbb{R})$ be a C^2-real function and consider the equation

$$(1.1) \qquad \ddot{x} = - V'(x,t)$$

where $x \in \mathbb{R}^N$, $\ddot{x} = \frac{d^2x}{dt^2}$ and V' denotes the gradient of V with respect to $x = (x_1,\ldots,x_N)$. We suppose that

V_1) V is $\tau = \frac{2\pi}{\Omega}$ periodic in t and $V(0,t) = 0, V'(0,t) = 0$ for any $t \in \mathbb{R}$

In this paper we are concerned with the existence of subharmonic solutions for (1.1) i.e. we search for $k\tau$-periodic solutions $(k \in \mathbb{N})x_k$ of (1.1). The research of subharmonic solutions of Hamiltonian systems is an old problem (cp. [7] and its references).

For example, under suitable assumptions on V near the origin and at infinity, there exists a sequence of $k\tau$-periodic solutions x_k whose minimal periods tend to ∞ as $k \to \infty$.

Nevertheless very little is known about the existence of subharmonic solutions of (1.1) with prescribed minimal period $k\tau$. In this paper we obtain some results in this direction in the case in which V is "superquadratic" at infinity. The "subquadratic" case has been studied in [6].

In order to state the results we need to introduce some notations.

Let

$$A = V''(0,t)$$

denote the hessian matrix of V at x = 0 and set[1]

$$\bar{V}(x,t) = V(x,t) - \frac{1}{2}(Ax|x) .$$

We assume that

V_2) $$A = \begin{pmatrix} \omega_1^2 & & 0 \\ & \ddots & \\ & & \omega_N^2 \end{pmatrix} \qquad 0 \leq \omega_1 \leq \dots \leq \omega_N$$

V_3) <u>There exists</u> $p > 2$ <u>and</u> $M > 0$ <u>s.t.</u>

$$(V'(x,t)|x) \geq p\, V(x,t) > 0 \quad \underline{\text{for}}\ |x| > M \quad \underline{\text{and}}\ t \in \mathbb{R}$$

V_4) $\bar{V}(x,t) \geq 0$ <u>for any</u> $x \in \mathbb{R}^N$ <u>and</u> $t \in \mathbb{R}$

V_5) V' <u>is odd</u> (i.e. $V'(-x,-t) = -V'(x,t)$)

V_6) $(V''(x,t)x|x) > (V'(x,t)|x)$
<u>for</u> $x \neq 0$ <u>and any</u> $t \in \mathbb{R}$

V_7) <u>If</u> $x = x(t)$ <u>is a periodic function with minimal period</u> $q\tau$, q <u>rational, and</u> $V'(x(t),t)$ <u>is periodic with minimal period</u> $q\tau$, <u>then</u> q <u>is integer</u>.

Observe that V_3) prescribes a superquadratic behaviour of V at infinity; in fact V_3) implies that

$$V(x,t) \geq \text{const}\ |x|^p \quad \text{for } |x| > M.$$

---

(1) $(\cdot|\cdot)$ denotes the inner product in $\mathbb{R}^N$.

Assumption V_7) is a generic one (cp. [6]); it is satisfied, for example, if

$$V(x,t) = g(t)\, U(x)$$

and g is a periodic function with minimal period τ .

The following theorem holds

Theorem 1.1. Let the assumptions V_1),...V_7) be satisfied. Let $k \in \mathbb{N}$ such that[2]

$$(1.2) \qquad k_o > \sum_{j=1}^{N} [k\, \omega_j/\Omega] + 1$$

where k_o is the least prime factor of k. Then (1.1) possesses a periodic solution with the minimal period $k\tau$.

The following corollaries are easily deduced from Theorem 1.1.

Corollary 1.2. Let the assumptions V_1),...,V_7) be satisfied and suppose that

$$\omega_1 = \omega_2 = \ldots = \omega_N = 0$$

Then for any $k \in \mathbb{N}$ (1.1) possesses a periodic solution with minimal period $k\,\tau$.

Corollary 1.3. Let the assumptions V_1),...,V_7) be satisfied. Let $k \in \mathbb{N}$ s.t.

$$(1.3) \qquad k < \frac{2\pi}{\omega_N} \qquad (\omega_N > 0)$$

---

(2) If $\alpha > 0$, $[\alpha]$ denotes the greatest integer $\leq \alpha$.

Then (1.1) *possesses a periodic solution with minimal period* $k\tau$.

Corollary 1.4. *Let the assumptions* $V_1),\dots,V_7)$ *be satisfied and suppose that*

$$(1.4) \qquad \delta = \sum_{j=1}^{N} \omega_j/\Omega < 1$$

then for any k *prime with*

$$k > \frac{1}{1-\delta}$$

(1) *possesses a periodic solution with minimal period* $k\tau$.

Corollary 1.5. *Let* $V_1),\dots,V_7)$ *be satisfied*. *Let* $p \in \mathbb{N}$, $p > 1$ *s.t.*

$$\frac{\omega_N}{\Omega} \le \frac{1}{p}$$

Then (1.1) *possesses* p-1 *periodic solutions with minimal period* τ, $2\tau,\dots, (p-1)\tau$.

Remark 1.6. Ambrosetti and Mancini have obtained in [1] results related to Corollary 1.3 for autonomous equations. Ekeland and Höfer in [3] and Girardi-Matzeu in [4] have obtained results related respectively to Corollary 1.2 and Corollary 1.3 for autonomous Hamiltonian systems. We recall that the results in [1,3,4] have been obtained under convexity assumptions on the potential V (or on the Hamiltonian function).

Remark 1.7. Observe that Corollary 1.4 permits to prove the existence of solutions of (1.1) with arbitrarily large minimal periods in some cases in which $A \equiv V''(0,t) \not\equiv 0$ and V is superquadratic at infinity.

2. SOME PRELIMINARIES

In this section we shall prove two results which will be useful in the proof of Theorem 1.1.

Let us first recall some definitions. Let f be a C^2-functional on a real Hilbert space E; we denote by f'(x) the Frechét differential of f at x and by f"(x) the bilinear form on $E \times E$ which represents the second differential of f at x. Suppose that

(2.1) <u>for any</u> $x \in E$ f"(x) <u>has discrete spectrum</u>.

If z is a critical point of f (i.e. f'(z) = 0) we denote by m(z) its Morse-index, i.e. we set

$$m(z) = \underline{\text{number of the negative eigenvalues (counted with}}$$
$$\underline{\text{their multiplicity}}\text{) of } f''(z)$$

Moreover we set

$$m^*(z) = m(z) + \dim \operatorname{Ker} f''(z) .$$

Assume that f satisfies the Palais-Smale condition, i.e.

(2.2) <u>Every sequence</u> $\{x_n\} \in E$ <u>s.t.</u> $\{f(x_n)\}$ <u>is bounded and</u> $f'(x_n) \to 0$ <u>contains a convergent subsequence</u>.

The following lemma permits to give, in some cases, an upper bound to the Morse index.

<u>Lemma 2.1</u>. <u>Let</u> f <u>be a</u> C^2<u>-functional on a real Hilbert space</u> E. <u>Suppose that</u> f <u>satisfies</u> (2.1), (2.2). <u>Suppose moreover that there exist con-</u>

stants ρ, R_1, R_2, $\alpha > 0$ with $R_1 > \rho$ and a finite-dimensional subspace V of E and $\phi \in V$, $||\phi|| = 1$ such that

$$f(u) \geq \alpha \qquad \forall\, u \in V, \ ||u|| = \rho$$

(2.3)

$$f(u) \leq 0 \qquad \forall\, u \in \partial Q$$

where ∂Q is the boundary of the Hilbert manifold

$$Q = \{v + t\phi : v \quad V, \ ||v|| \leq R_2, \ t \in [0, R_1]\}$$

Under the above assumptions f possesses a critical point u with Morse index

(2.4) $\qquad m(u) \leq \dim V + 1$.

Moreover

(2.5) $\qquad \alpha \leq f(u) \leq \sup f(Q) \equiv \beta$

proof.

First introduce some notations:

If $c, d \in \mathbb{R}$ with $c < d$ we set

$$\Delta^c = \{x \in E \mid f(x) \leq c\}$$

$$\Delta_c^d = \{x \in E \mid c \leq f(x) \leq d\} .$$

We denote by $H_*(\cdot,\cdot)$ the relative singular homology with coefficients in some field F.

Now let $\varepsilon > 0$ such that $\alpha - \varepsilon > 0$. By the Sard lemma in infinite dimensional spaces, there exist two regular values a,b s.t.

$$\alpha - \varepsilon < a < \alpha \qquad \text{and} \qquad \beta < b < \beta + \varepsilon$$

We recall that

$$S = \{x \in V^{\perp}, \mid ||x|| = \rho\} \quad \text{and} \quad \partial Q$$

satisfy "the linking property" (e.g. cp [2, prop. 2.2]); then, by using (2.3), we deduce that Q is a cycle which is not a boundary in Δ^a; then, if $q = \dim V$, we deduce that

(2.6) $$\gamma \equiv [\partial Q] \in H_q(\Delta^a) \;, \quad \gamma \neq 0 \;.$$

Now consider the map

$$i^* : H_q(\Delta^a) \to H_q(\Delta^b) \qquad (q = \dim V)$$

induced by the canonical embedding

$$i : \Delta^a \to \Delta^b \;.$$

Since ∂Q is the boundary of $Q \subset \Delta^b$ we have

(2.7) $$\gamma \in \operatorname{Ker} i^* \;.$$

Then, by using the exactness of the sequence

$$\ldots \to H_{q+1}(\Delta^b, \Delta^a) \overset{\partial^*}{\to} H_q(\Delta^a) \overset{i^*}{\to} H_q(\Delta^b) \quad \ldots$$

we deduce that

(2.8) $$\gamma \in \operatorname{Ker} i^* = \operatorname{Im} \partial^*$$

Then, by (2.6) and (2.8), we conclude that

(2.9) $$H_{q+1}(\Delta^b, \Delta^a) \neq 0 \qquad\qquad (q = \dim V)$$

Now set

(2.10) $$i(\Delta_a^b) = \sum_{\ell=0}^{\infty} \dim H_\ell(\Delta^b,\Delta^a)\, t^\ell$$

Then, since $H_{q+1}(\Delta^b,\Delta^a) \neq 0$, we deduce that

(2.11) $$i(\Delta_a^b) = t^{q+1} + z(t)$$

where z(t) is a formal series with positive coefficients. It can be shown (cp. Benci in this volume) that

(2.12) $$i(\Delta_a^b) = \sum_{\ell=m(K)}^{m^*(K)} c_\ell\, t^\ell \qquad c_\ell > 0$$ (3)

where

$$K = \{x \in E \mid f'(x) = 0, \quad a \leq f(x) \leq b\}$$

and

$$m(K) = \inf\{m(x) : x \in K\}$$

$$m^*(K) = \sup\{m^*(x) : x \in K\} .$$

From (2.11), (2.12) we deduce that there exists $x \in K$ s.t. $m(x) \leq q+1$.

Remark 2.2. The existence of a critical point satisfying (2.5) can be proved, under the assumptions of lemma 2.1, by using min-max arguments (cp. e.g. [2] and its references). Nevertheless sometimes it is useful (as we shall see) to have "informations", like the bound (2.4), on the Morse-index of the critical point. We recall that a result related to

---

(3) In the conference of Benci $i(\Delta_a^b)$ is defined by using the Alexander-Spanier Cohomology. Nevertheless (2.12) holds also if we define $i(\Delta_a^b)$ as in (2.10). In fact, in this case, we have

$$\dim H^\ell(\Delta^b,\Delta^a) = \dim H_\ell(\Delta^b,\Delta^a)$$

where $H^*(\cdot,\cdot)$ denotes the Alexander-Spanier Cohomology.

to lemma 2.1 has been proved bay Lazer and Solimini [5].

The following lemma is useful to get, in some situations, a lower bound to the Morse-index.

Lemma 2.3. Suppose that the potential function $V = V(x,t)$ *satisfies assumptions* V_6) (cp. *Theorem* 1.1) *and let* $z = z(t)$ *be a nontrivial solution of the boundary value problem.*

$$(2.13) \qquad \begin{aligned} &\ddot{x} + V'(x,t) = 0 \quad \text{in }]0,T[\ , \quad T > 0 \\ &x(0) = x(T) = 0 \ . \end{aligned}$$

Suppose moreover that there exist points

$$T_0 = 0 < T_1 < \ldots < T_{\ell-1} < T_\ell = T$$

s.t. $\qquad z(T_i) = 0 \qquad\qquad i = 0,\ldots,\ell$

Then

$$\ell \leq m(z)$$

where $m(z)$ *denotes the Morse-index of* z.

Proof. Set

$$\tau_i =]T_{i-1}, T_i[\qquad\qquad i = 1,\ldots,\ell$$

and

$$\alpha_i(t) = \begin{cases} z(t) & \text{if } t \quad \tau_i \\ 0 & \text{if } t \quad \tau_i \end{cases} \qquad\qquad i = 1,\ldots,\ell$$

Obviously α_i $(i = 1,\ldots,\ell)$ are linearly independent.

Let V_ℓ denote the vector space spanned by $\{\alpha_i\}$ and consider the functional whose critical points are the solutions of (2.13)

$$f(x) = \int_0^T \left(\frac{1}{2}\,|\dot{x}(t)|^2 \ - \ V(x(t),t)\right) dt \qquad x \in H_0^1(]0,T[).$$

We want to prove that $f''(z)$ is negative-definite on V_ℓ. Let

$$v \in V_\ell \setminus \{0\} \qquad v = \sum_{i=1}^{\ell} c_i \alpha_i \qquad c_i \in \mathbb{R}$$

then

$$<f''(z)v,v> = \int_0^T (|\dot{v}|^2 - (V''(z,t)v|v))\, dt =$$

$$= \sum_{i=1}^{\ell} \int_{\tau_i} (|\dot{v}|^2 - (V''(z,t)v|v))\, dt =$$

$$= \sum_{i=1}^{\ell} c_i^2 \int_{\tau_i} (|\dot{z}|^2 - (V''(z,t)z|z))\, dt < (\text{by } V_6) <$$

$$< \sum_{i=1}^{\ell} c_i^2 \int_{\tau_i} (|\dot{z}|^2 - (V'(z,t)|z))\, dt = 0$$

The last equality follows from the fact that $z = z(t)$ solves (1.1) in τ_i and $z(T_{i-1}) = z(T_i) = 0$ $(i = 1,\dots,\ell)$.

3. PROOF OF THEOREM 1.1

We are now ready to prove Theorem 1.1.

Consider a positive integer k which satisfies (1.2). Set

$$2T = k\,\tau = k\frac{2\pi}{\Omega}\,.$$

Consider the boundary value problem

$$(3.1)\qquad \begin{cases} \ddot{x} + V'(x,t) = 0 & \text{in }]0,T[\\ x(0) = x(T) = 0\,. \end{cases}$$

The solutions of (3.1) are the critical points of the functional

$$f(x) = \int_0^T (\tfrac{1}{2}(|\dot{x}|^2 - V(x,t))\, dt \qquad x \in H_0^1(]0,T[)$$

Obviously

$$f(x) = \frac{1}{2} (L x|x) - \int_0^T \hat{V}(x,t)\, dt$$

where L is the self-adjoint realization in $L^2(]0,T[)$ of the operator $x \to -\ddot{x} - A x$ with boundary condition $x(0) = x(T) = 0$ and $(\cdot|\cdot)$ denotes the inner product in $L^2(]0,T[)$.

We shall use lemma 2.1. By assumption V_3) standard calculations show that f satisfies the Palais-Smale condition (2.2).

The eigenfunctions of L are

$$\phi_{jn} = e_j \sin n \frac{\pi}{T} t \qquad j = 1,\dots,N \qquad n \in \mathbb{N} \setminus \{0\}$$

where $\{e_j\}$ is the canonical base in $\mathbb{R}^N$.

The corresponding eigenvalues are

(3.2) $$\lambda_{jn} = (n \frac{\pi}{T})^2 - \omega_j^2 \qquad j = 1,\dots,N \qquad n \in \mathbb{N} \setminus \{0\}$$

where ω_j^2 $(j = 1,\dots,N)$ are the eigenvalues of A.

Set

$$H_+ = \overline{\text{span} \{ \phi_{jn} : \lambda_{jn} > 0 \}} , \qquad H_- = H_+^\perp$$

where the closure is taken in $H_0^1(]0,T[)$.

Since

$$\hat{V}(x,t) = o(|x|^2) \quad \text{at} \quad x = 0$$

standard calculations show that there exist $\rho, \alpha > 0$ s.t.

(3.3) $$f(x) \geq \alpha \qquad \text{for} \qquad x \in H_+ \quad ||x|| = \rho$$

Let R_1, R_2 be two positive numbers with $R_1 > \rho$ and consider $\Phi \in H_+$ with $||\Phi|| = 1$.

Set

$$Q \equiv \{v + t\phi \mid v \in H_-, \ ||v|| \leq R_2, \ t \in [0,R_1]\}.$$

By V_4) we immediately deduce that

(3.4) $\quad f(x) \leq 0 \qquad \forall\, x \in H_-$.

Moreover, by the superquadratic growth of V at ∞ (cf. assumption V_3), standard calculations show that

(3.5) $\quad \lim\limits_{||x|| \to +\infty} f(x) = -\infty \qquad \text{for } x \in H_- \oplus \{t\phi\}_{t \in \mathbb{R}}$

Then, if R_1 and R_2 are sufficiently large, from (3.4) and (3.5) we easily deduce that

(3.6) $\quad f(x) \leq 0 \qquad \forall\, x \in \partial Q$

By virtue of (3.6) and (3.3), also assumption (2.3) of lemma 2.1 is satisfied. Then, by lemma 2.1, (3.1) possesses a non-trivial solution $x = x(t)$ with Morse index

(3.7) $\quad m(x) \leq \dim H_- + 1.$

Now, by using (3.2), we deduce that

(3.8) $\quad \dim H_- = \#\{(n,j) : \frac{n\pi}{T} \leq \omega_j;\ j = 1,\dots,N;\ n \in \mathbb{N} \setminus \{0\}\} =$

$$= \sum_j \left[\frac{T\omega_j}{\pi}\right] = \sum_j \left[\frac{k\omega_j}{\Omega}\right] \qquad \left(T = \frac{k\tau}{2} = \frac{k\pi}{\Omega}\right)$$

Consider now the function $\bar{x}(t)$ obtained by expanding $x(t)$ by oddness to a 2T-periodic function (i.e. $\bar{x}$ is the 2T-periodic function s.t. $\bar{x}(t) = x(t)$ if $t \in [0,T]$ and $\bar{x}(t+T) = - x(T-t)$ if $t \in [0,T]$).

Since V' is odd, $\bar{x}$ is a 2T-periodic solution of (1.1). Now we prove that $\bar{x}$ has minimal period $k\tau$.

Arguing by contradiction we suppose that $\bar{x}$ has not minimal period $k\tau$. Then, by V_7), $\bar{x}$ has minimal period $\leq k\tau/k_o$, where k_o is the least prime factor of k. Consequently there exist points $T_o \equiv 0 < T_1 < \ldots < T_{k_o-1} < T \equiv T_{k_o}$ such that $x(T_i) = 0$ ($i = 0,\ldots,k_o$).

Then, by using lemma 2.3, we get

$$k_o \leq m(x) \tag{3.9}$$

By (3.7), (3.8), (3.9) we deduce

$$k_o \leq \sum_j \left[\frac{k\,\omega_j}{\Omega}\right] + 1$$

which contradicts (1.2).

REFERENCES

[1] Ambrosetti, A., Mancini, G.: "Solutions of minimal period for a class of convex Hamiltonian systems", Math. Ann. 255, 405-421 (1981)

[2] Bartolo, P., Benci, V., Fortunato, D.: "Abstract critical point theorems and applications to some nonlinear problems with "strong" resonance at infinity", Nonlinear Anal. Theory, Methods and Appl. 7, 981-1012 (1983)

[3] Ekeland, I., Höfer, H.: "Periodic solution with prescribed period for convex autonomous hamiltonian systems, preprint CEREMADE n. 8421, Paris (1984)

[4] Girardi, M., Matzeu, M.: "Periodic Solutions of Convex autonomous hamiltonian systems with a quadratic growth at the origin and superquadratic at infinity, preprint (1985)

[5] Lazer, A., Solimini, S.: "Nontrivial solutions of operator equations and Morse indices of critical points of min-max type, preprint (1985)

[6] Michalek, R., Tarantello, G.: "Subharmonic solutions with prescribed minimal period for non-autonomous Hamiltonian systems" preprint (1986)

[7] Rabinowitz, P.H.: "On subharmonic solutions of hamiltonian systems", Comm. Pure Appl. Math. 33, 609-633 (1980).

EXAMPLES OF LONG TIME SCALES IN HAMILTONIAN DYNAMICAL SYSTEMS

Giancarlo Benettin

Dipartimento di Fisica dell'Università di Padova
Via F. Marzolo 8 - 35131 Padova
ITALY

1. NEKHOROSHEV EXPONENTIAL ESTIMATES

The purpose of this talk is to illustrate some basic aspects of classical perturbation theory for Hamiltonian dynamical systems of the form

$$H_\varepsilon(p,q) = h(p,q) + \varepsilon f(p,q) \ , \tag{1}$$

where $(p,q) = (p_1,\dots,p_n,q_1,\dots,q_n)$ is any set of canonical coordinates, say $(p,q) \in \mathcal{B} \subset \mathbb{R}^{2n}$, and ε is a small parameter. Given ε and, say, some general information on f, like a convenient norm, one wants to know at which time scale the effects of the perturbation εf, added to the unperturbed Hamiltonian h, are expected to become relevant.

The answer clearly depends on what is considered to be relevant. If one is interested in orbits, then the only general result one can produce is the uniqueness theorem for the solution of ordinary differential equations: denoting by $\Phi_\varepsilon^t(p,q)$ the solution of Hamilton equations corresponding to Hamiltonian (1) with initial datum (p,q), one has clearly $\Phi_\varepsilon^t(p,q) \to \Phi_0^t(p,q)$ for $\varepsilon \to 0$. However, the limit is highly non-uniform in time, as one can only guarantee very poor estimates of the form

(2) $$\operatorname{dist}(\Phi_\varepsilon^t(p,q),\Phi_\circ^t(p,q)) \le K\varepsilon e^{\chi t},$$

K being a constant independent of ε and t. Such an estimate is, unfortunately, optimal whenever the unperturbed system has unstable orbits; it shows that informations on orbits are rapidly lost, on a time scale $t \sim \chi^{-1}\log(K\varepsilon)^{-1}$.

The situation is quite different if the term "relevant" is referred to the behaviour of the integrals of motion of h. Indeed, suppose there exists a function $J : \mathcal{B} \to \mathbf{R}$, such that the Poisson bracket $\{h,J\}$ vanishes. Then one has $\partial J/\partial t = \varepsilon\{f,J\}$, so that one obtains estimates of the form

(3) $$|J(\Phi_\varepsilon^t(p,q)) - J(p,q)| \le K\varepsilon t ,$$

i.e., we are now loosing informations on the value of J on a longer time scale, $t \sim (K\varepsilon)^{-1}$.

The basic purpose of classical perturbation theory is to go beyond this zeroth-order result. This can be done by introducing a canonical change of variables $(p,q) = \mathcal{C}_\varepsilon^{(r)}(p',q')$, $\mathcal{C}_\varepsilon^{(r)} : \mathcal{B}' \to \mathcal{B}$ ε-close to the identity, such that the new Hamiltonian $H'(p',q') \equiv H_\varepsilon \circ \mathcal{C}_\varepsilon^{(r)}$ has the form

(4) $$H_\varepsilon'(p',q') = h(p',q') + \sum_{s=1}^{r} h_s(p',q') + \varepsilon^{r+1}f_r'(p',q',\varepsilon).$$

If one is able to impose $\{h_s,J\} = 0$, $s = 1,\dots,r$, then one gets the estimate of order r

(5) $$|J(\Phi_\varepsilon'^t(p',q')) - J(p',q')| \le K_r\varepsilon^{r+1}t ,$$

where $\Phi_\varepsilon'^t = (\mathcal{C}_\varepsilon^{(r)})^{-1} \circ \Phi_\varepsilon^t \circ \mathcal{C}_\varepsilon^{(r)}$ is the solution of the Hamiltonian problem (4). It also follows, for the old dynamics, an estimate of the form

$$(6) \qquad |J(\Phi_\varepsilon^t(p,q)) - J(p,q)| \le K_r \varepsilon^{r+1} t + C(\varepsilon) \; ,$$

$C(\varepsilon)$ being small (and independent of t) for a canonical transformation close to the identity. Estimates (5), (6) show that our control on the integral of motion J is now lost after a time scale $t \sim K_r^{-1} \varepsilon^{-r-1}$.

In order to accomplish this program, one is basically confronted with two problems (see ref. [1] for a deeper insight):

1. A feasibility problem: indeed, it is non obvious that the canonical transformation $\mathcal{C}_\varepsilon^{(r)}$ does exist. As is well known, its generating function is required to satisfy, at each order in ε, a partial differential equation, which in general (as first understood by Poincaré [2]) does not admit solution, unless one introduces special assumptions or suitable tricks.
2. A convenience problem, depending on the growth rate of K, with r. Typically one finds estimates of the form $K_r \le (Ar^a)^r K_o$ (A, a and K_o being positive constants) so that for any fixed ε, no matter how small, convergence of the perturbative series is not found. Consequently, for each finite ε there will be a finite value of r at which it is most convenient to stop perturbation theory.

Let us first concentrate on this latter problem. The point is that even the above apparently poor estimate for K_r leads to nice results: indeed, if one makes the choice $r = (\varepsilon_o/\varepsilon)^{1/a}$, with $\varepsilon_o = (Ae)^{-1}$ then from (5) one immediately obtains the exponential estimate

$$(7) \qquad |J_\varepsilon^{(r)}(\Phi_\varepsilon^t(p,q)) - J_\varepsilon^{(r)}(p,q)| \le K_o \varepsilon t e^{-(\varepsilon_o/\varepsilon)^{1/a}} ,$$

which leads to a control on the system for the "long time scale" $t \sim e^{(\varepsilon_0/\varepsilon)^{1/a}}$. This consideration is fully elementary, but is the kernel of the celebrated Nekhoroshev theorem [3]; it shows that, once feasibility problems are solved, then the exponential estimates naturally follow. Such exponential estimates are almost optimal: examples are known [4] of non-perturbative phenomena ("Arnold diffusion") which take place on an exponential time scale $t \sim e^{(\varepsilon_0/\varepsilon)^{\frac{1}{2}}}$; unfortunately, Nekhoroshev theorem gives in general $a \gg 2$.

Let us turn to the more serious question of feasibility, and show the main difficulty within the simplest possible framework, i.e., for analytic perturbations of a system of harmonic oscillators. Using the action-angle coordinates, the Hamiltonian is written

$$(8) \qquad H_\varepsilon = \sum_{i=1}^{n} \omega_i J_i + \varepsilon f(J,\phi),$$

with $J = (J_1,\ldots,J_n) \in \quad \times R^n$, $\phi = (\phi_1,\ldots,\phi_n) \in T^n$; f is assumed to be analytic in $D = \quad \times T^n$. The canonical transformation $C_\varepsilon^{(r)}$ can be written in the form

$$(9) \qquad J = J' + \frac{\partial S}{\partial \phi}, \qquad \phi' = \phi + \frac{\partial S}{\partial J'},$$

$S(J',\phi) = \varepsilon S_1(J',\phi) + \varepsilon^2 S_2(J',\phi) + \ldots + \varepsilon^r S_r(J',\phi)$ being a generating function to be determined.

We may want to obtain the new Hamiltonian in the form

$$(10) \qquad H'(J',\phi') = \sum_{i=1}^{n} \omega_i J_i + \sum_{s=1}^{r} \varepsilon^s h_s(J') + \varepsilon^{r+1} f_r(J',\phi'),$$

in order it has, up to order r, just n integrals of motion.

Let us see how one determins S_1. One must replace (9) in (8), and eliminate the ϕ-dependence at first order in ε. This leads to the equation

$$(11) \qquad \sum_{i=1}^{n} \omega_i \frac{\partial S_1}{\partial \phi_i}(J',\phi) + f(J',\phi) = h_1(J') ,$$

where both S_1 and h_1 are unknown. Clearly, h_1 must be the average of f over the angles, while the Fourier components $\tilde{S}_k$, $\tilde{f}_k$ of S and f, for nonvanishing $k \in Z^n$, must be related by

$$(12) \qquad (ik \cdot \omega)\tilde{S}_k(J') + \tilde{f}_k(J') = 0$$

(the dot denotes here the ordinary scalar product). This equation cannot be solved by

$$(13) \qquad \tilde{S}_k = - \frac{\tilde{f}_k}{ik \cdot \omega} ,$$

unless one has $k \cdot \omega \neq 0 \quad \forall\, k \quad Z^n$. In fact, to assure convergence of the Fourier series to a regular function, one needs more, precisely the celebrated Diophantine condition

$$(14) \qquad |k \cdot \omega| \geq \gamma |k|^{-n}, \qquad \gamma > 0, \qquad |k| \equiv |k_1| + \ldots + |k_n| ,$$

which is satisfied by suitable γ for almost all ω's, although, for any finite γ, a dense open set is excluded.

It could be seen that higher order terms of the generating function must satisfy equations similar to (11), with a suitable known term in place of f, so that new difficulties are not produced, although a lot of annoying technical estimates cannot be avoided. As a result one obtains the following version of Nekhoroshev theorem, due to G. Gallavotti [5] [6] ($\| \cdot \|_D$ denotes here the supremum norm in D):

Theorem 1: Consider the Hamiltonian

(15) $$H = \sum_{i=1}^{n} \omega_i J_i + \varepsilon f(J,\phi) ,$$

and assume:

1. f is analytic in $\mathcal{D}_{R,\rho,\xi}$ given by

(16) $$|\mathrm{Re} J_i| \leq R + \rho , \quad |\mathrm{Im} J_i| \leq \rho , \quad |\mathrm{Im}\phi| \leq \xi \leq 1 ,$$

and satisfies (this is not restrictive, and gives a precise meaning to ε)

(17) $$\max_{1 \leq i \leq n} \left\| \frac{\partial f}{\partial J_i} \right\|_{\mathcal{D}_{R,\rho,\xi}} , \frac{1}{\rho} \left\| \frac{\partial f}{\partial \phi_i} \right\|_{\mathcal{D}_{R,\rho,\xi}} = \max_{1 \leq i \leq n} |\omega_i| \equiv F .$$

2. satisfies the Diophantine condition

(18) $$|\omega \cdot k| \geq \gamma |k|^{-n} \quad \forall\, k \quad Z^n, \quad k \neq 0 ;$$

3. is small, precisely,

(19) $$\varepsilon \leq \varepsilon_o(n,\xi,F/\gamma) \equiv C_n \xi^{-4n-2} (F/\gamma)^{-2} ,$$

where C_n is a constant depending only on n.

Then there exists a real analytic canonical transformation $(J,\phi) = \mathcal{C}_\varepsilon(J',\phi')$, $\mathcal{C}_\varepsilon : \mathcal{D}_{R,\frac{1}{2}\rho,\frac{1}{2}\xi} \to \mathcal{D}(R,\rho,\xi)$, satisfying

(20) $$|J_i' - J_i| < \rho \left(\frac{\varepsilon}{\varepsilon_o}\right)^{\frac{1}{2}} , \quad |\phi_i' - \phi_i| < \xi \left(\frac{\varepsilon}{\varepsilon_o}\right)^{\frac{1}{2}} , \quad 1 \leq i \leq n ,$$

which gives the new Hamiltonian $H_\varepsilon' \equiv H_\varepsilon \circ \mathcal{C}_\varepsilon$ the form

$$(21) \qquad H'_\varepsilon = \omega\cdot J' + \varepsilon g(J',\varepsilon) + \varepsilon\, e^{-(\varepsilon_0/\varepsilon)^{1/a}} f'(J',\phi',\varepsilon) ,$$

with $a = 4(n+1)$, and

$$(22) \qquad \max_{1\le i\le n} \left\| \frac{\partial f'}{\partial J_i} \right\|_{\mathcal{D}_{R,\frac{1}{2}\rho,\frac{1}{2}\xi}}, \frac{1}{\rho}\left\| \frac{\partial f'}{\partial \phi_i} \right\|_{\mathcal{D}_{R,\frac{1}{2}\rho,\frac{1}{2}\xi}}, \frac{1}{2}\left\| \frac{\partial g}{\partial J_i} \right\|_{\mathcal{D}_{R,\frac{1}{2}\rho,\frac{1}{2}\xi}} < F,$$

so that one has in particular

$$(23) \qquad \begin{gathered} |J'_i(t) - J'_i(0)| \le \varepsilon\rho, \quad |J_i(t) - J_i(0)| \le 2\rho \left(\frac{\varepsilon}{\varepsilon_0}\right)^{\frac{1}{2}}, \quad 1\le i\le n, \\ \text{for} \quad |t| \le F^{-1}\, e^{(\varepsilon_0/\varepsilon)^{1/a}} \end{gathered}$$

For other Nekhoroshev-like results, in different and more general cases, see Nekhoroshev papers [3], or ref. [1] [6] [7].

2. LONG TIME SCALES AND REALIZATION OF HOLONOMIC CONSTRAINTS

2.1 Let us consider a Lagrangian dynamical systems with $n + \nu$ degrees of freedom, and suppose it reduces to n degrees of freedom via the introduction of ν holonomic constraints. Suppose that the $n+\nu$ Lagrangian coordinates, denoted $(x_1,\ldots,x_n,\xi_1,\ldots,\xi_\nu)$, can be chosen in such a way that $\xi_1 = 0,\ldots,\xi_\nu = 0$ are the equations of the constraint, and consider the full expression of the kinetic energy:

$$(24) \qquad T = \frac{1}{2}\sum_{i,j=1}^{n} A_{ij}(x,\ )\dot{x}_i\dot{x}_j + \frac{1}{2}\sum_{i,j=1}^{\nu} B_{ij}(x,\xi)\dot{\xi}_i\dot{\xi}_j + \sum_{i=1}^{n}\sum_{j=1}^{\nu} C_{ij}(x,\xi)\dot{x}_i\dot{\xi}_j$$

As shown in ref. [8], it is not restrictive to assume

$$(25) \qquad \begin{aligned} A_{ij}(x,\xi) &= a_{ij}(x) + \mathcal{O}(\xi) \\ B_{ij}(x,\xi) &= b_{ij} + \mathcal{O}(\xi) \\ C_{ij}(x,\xi) &= \mathcal{O}(\xi) \end{aligned}$$

(the first equality is trivial, the last expresses the orthogonality of the x and ξ coordinates on the constraint, the second is obtained via a convenient choice of the metrics on the constraint).

One has then

(26) $$T = T^{(0)}(x,\dot{x}) + T^{(1)}(\dot{\xi}) + O(\xi),$$

with $T^{(0)} = \frac{1}{2} \Sigma\, a_{ij}(x), \dot{x}_i, \dot{x}_j$, $T^{(1)} = \frac{1}{2} \Sigma\, b_{ij} \dot{\xi}_i \dot{\xi}_j$.

If $V(x,\xi) = V^{(0)}(x) + O(\xi)$ is the potential energy, which is assumed to be bounded from below on the constraint, then the constrained motion is given by the Lagrangian

(27) $$L^{(0)}(x,\dot{x}) = T^{(0)}(x,\dot{x}) - V^{(0)}(x)$$

The problem of the "realization of constraints" is the study of a physical model for constraints: instead of reducing from n+ to n the number of degrees of freedom by directly eliminating the variables ξ, in a purely geometrical way, one considers the dynamics for the whole set of $n+\nu$ variables, and represents the constraint by an additional "confining potential" $\omega^2 W(x,\xi)$, satisfying $W(x,0) = 0$ and $W(x,\xi) \neq 0$ for $\xi \neq 0$. The physical intuition is that in the limit $\omega \to \infty$ one should obtain the constrained motion. However, the question is definitely non-trivial, becasue the above potential give spurely positional forces, while constraint forces are well known to be velocity dependent (and also dependent on the choice of V).

Let

(28) $$L_\omega(x,\dot{x},\xi,\dot{\xi}) = T(x,\dot{x},\xi,\dot{\xi}) - V(x,\xi) - \omega^2 W(x,\xi)$$

be the full Lagrangian, and denote by $(X_\omega(t,x,\dot{x},\xi,\dot{\xi}), \Xi_\omega(t,x,\dot{x},\xi,\dot{\xi}))$ the solution of the equation of motion with initial datum $(x,\dot{x},\xi,\dot{\xi})$. The question is then whether $X_\omega(t,x,\dot{x},0,\dot{\xi})$ has a limit $X_\infty(t,x,\dot{x},\dot{\xi})$ for

$\omega \to \infty$ at fixed t, and whether in addition $X_\infty(t,x,\dot{x},\dot{\xi})$ is independent of $\dot{\xi}$, and coincides with the solution $X^{(0)}(t,x,\dot{x})$ of the constrained Lagrangian problem (27). Ξ_ω is not required to have a limit; ξ-variables are expected to oscillate faster and faster, and can save in the limit a nonvanishing amount of energy.

A sufficient condition for a positive answer to both questions is given by the following theorem, which is due to Rubin and Hungar [9], Gallavotti [8], and Tackens [10]:

Theorem 2: Within the above conditions, if the confining potential $W(x,\xi)$ is independent of $x, W(x,\xi) = \mathcal{W}(\xi)$, then for any fixed t one has

$$\lim_{\omega \to \infty} X_\omega(t,x,\dot{x},0,\dot{\xi}) = X^{(0)}(t,x,\dot{x}) \tag{29}$$

The proof is easy, and basically reduces to the uniqueness theorem for the solution of ordinary differential equations. Unfortunately as in that case, the only general estimates one is able to produce have the form

$$\operatorname{dist}(X_\omega(t,x,0,\dot{\xi}),X^{(0)}(t,x,\dot{x})) \leq \omega^{-1} K e^{\chi t} , \tag{30}$$

and are optimal whenever the system one wants to realize has unstable orbits, so that they turn out to be completely unuseful after a time scale of order $\log \omega$. Here too, however, we can assume the same attitude we had when describing perturbation theory, i.e. we can look at the integrals of motions of the system we want to realize, first of all at its energy.

2.2 Let us switch to the Hamilton formalism, considering for the moment the simpler case $\nu = 1$ (but any n); we also assume $\mathcal{W}(\xi) = \frac{1}{2}\xi^2$

(however, this latter restriction is not essential). Denote by $\pi, p_1, \ldots, p_n$ the conjugate momenta to $\xi, x_1, \ldots, x_n$; from (28), taking into account (26), (27), we obtain for the full Hamiltonian the form

$$(31) \qquad H_\omega = \frac{1}{2}\pi^2 + h(p,x) + \xi U(p,x,\pi,\xi) + \frac{1}{2}\omega^2\xi^2 \quad ,$$

where $h(p,x)$ is the Hamiltonian of the constrained problem, while U is regular and bounded in a natural domain of definition, say for given finite total energy.

Let us introduce the action-angle coordinates (J,ϕ) by

$$(32) \qquad \pi = \sqrt{2\omega J}\cos\phi, \qquad \xi = \omega^{-1}\sqrt{2\omega J}\sin\phi \ .$$

The Hamiltonian assumes the form

$$(33) \qquad H_\omega = \omega J + h(p,x) + \omega^{-1} f(p,x,\omega J,\phi,\omega) \ ,$$

f being bounded whenever the energy ωJ of the motion "transversal to the constraint" is bounded. Thus, it will be natural to consider the Hamiltonian (33) in a domain D of the form

$$(34) \qquad (p,x) \in B \times R^n \ , \qquad \omega J \leq E_\circ \ , \qquad \phi \in T^n \ .$$

Denote by $E = \omega J$ the energy of the vibrational motion orthogonal to the constraint. From (33) it is clear that sensible energy exchanges between h and E require at least a time scale of order ω^{-1}, but it is quite natural to try to improve this zeroth-order result by means of classical perturbation theory. The obvious idea is to search for a canonical transformation $(p,x,J,\phi) = C^r_\omega(p',x',J',\phi')$ such that the new Hamiltonian $H^{(r)} \equiv H \circ C^r_\omega$ has the form

(35) $$H_\omega^{(r)} = \omega J' + (h(p',x') + \omega^{-1} g^{(r)}(p',x',\omega J',\omega) + \omega^{-r-1} f^{(r)}(p',x',\omega J',\omega)$$

indeed, in this case we obtain for $E' = \omega J'$ an estimate of the form

(36) $$|E'(t) - E'(0)| \leq K_r \omega^{-r} t,$$

which, as we have seen in the previous section, easily turns into an exponential estimate as far as one is able to keep an even poor control on K_r.

Before giving a precise statement, let us just sketch how perturbation theory does work with a Hamiltonian like (33); to this purpose it is sufficient to consider the simple case r = 1. Among the possible ways to introduce canonical transformations close to the identity, let us use here the Lie method, which has several advantages in particular if one wants to optimize the dependence of the estimates on the number of degrees of freedom.

Let us recall the essence of the Lie method. $\mathcal{C}_\omega(p',x',J',\phi')$ is defined as the time-one map of an auxiliary Hamiltonian problem, with unknown Hamiltonian $S_\omega(p,x,J,\phi)$ and initial datum (p',x',J',ϕ'). Any function F turns then into $F \circ \mathcal{C}_\omega = F + \{S_\omega, F\} + \mathcal{O}(S_\omega^2)$ (formally, $e^{\{S_\omega,.\}}F$). For the Hamiltonian (33) one finds in particular

(37) $$\begin{aligned} H_\omega \circ \mathcal{C}_\omega = {} & \omega J' + h + \omega^{-1} f \\ & + \{S_\omega, \omega J\} + \{S_\omega, h\} + \{S_\omega, \omega^{-1} f\} + \mathcal{O}(S_\omega^2). \end{aligned}$$

In order to eliminate the ϕ-dependent terms of order ω^{-1}, we impose that $\{S_\omega, \omega J\} + f$ is ϕ-independent: we thus require

(38) $$-\omega \frac{\partial S_\omega}{\partial \phi} + \omega^{-1} f = \omega^{-1} \bar{f},$$

where $\bar{f}(p,x,\omega J,\omega)$ is the average of $f(p,x,\omega J,\phi,\omega)$ over ϕ. This equation is solved by

(39) $$S_\omega = \omega^{-2} \int^{\phi}(f - \overline{f})d\phi'$$

(notice that no "small denominator" is present). The second line of (37) is now of order ω^{-2}, although this is not as trivial as it can appear: the point if that f, and consequently S_ω, are supposed to be regular functions not of J but of the product ωJ, so that J-derivatives acquire an extra factor ω. However, fortunately h does not depend on (J,ϕ), while f has in front the compensating factor ω^{-1}. Similar considerations show that the $O(S_\omega^2)$ term also has a leading contribution of order ω^2.

A simple iteration of this procedure leads formally to (35), with no complications at all. Moreover, in the analytic case the estimates are quite easy, and lead to the exponential estimates, as is guaranted by the following theorem [11]:

Theorem 3: Consider the real domain D given by

(40) $$(p,q) \in B \subset R^{2n}, \quad \frac{1}{2} E_0 \leq \omega J \leq E_0, \qquad \phi \in T^1 ,$$

and its complex extension D_ρ given by

(41) $$\{(p,x,J,\phi) \in C^{2n-2}; \quad (\tilde{p},\tilde{x},\tilde{J},\tilde{\phi}) \in D : \quad |p_i - \tilde{p}_i|, |x_i - \tilde{x}_i| \leq \rho, \ 1 \leq i \leq n; \ |\omega J - \omega\tilde{J}|, |\phi - \tilde{\phi}| \leq \rho\}$$

1. Assume that the Hamiltonian

(42) $$H_\omega = \omega J + h(p,x) + \omega^{-1} f(p,x,\omega J,\phi,\omega)$$

is analytic in the interior of D_ρ, and denote

(43) $$F = \max(\| H \|_\rho, \| h \|_\rho, \| f \|_\rho) ,$$

where $\| . \|_\rho$ is the supremum norm in D_ρ.

2. Assume $F/\rho \geq 1$, and $\omega \geq \omega_o(F,\rho) \equiv K(F/\rho)^2$, K being a convenient purely numeric constant larger than one (a possible value is 2^{16}).

Then there exists a real analytic canonical transformation $(p,x,J,\phi) = C_\omega(p',x',J',\phi')$, $C_\omega : D_{\frac{1}{2}\rho} \to D_\rho$, $C_\omega(D_{\frac{1}{2}\rho}) \supset D$, close to the identity in the sense that for $(p,x,J,\phi) \in D$ one has

(44) $$\begin{aligned} |p_i' - p_i|, \ |x_i' - x_i| &< \omega^{-1} F/\rho , \qquad 1 \leq i \leq n \\ |\omega J' - \omega J| &< 2\omega^{-1} F/\rho, \\ |\phi' - \phi| &< 2^5 \omega^{-1} F/\rho, \end{aligned}$$

which gives the new Hamiltonian $H' \equiv H \circ C_\omega$ the form

(45) $$H' = \omega J' + h(p',x') + \omega^{-1} g(p',x',\omega J',\omega) + \omega^{-2} e^{-(\omega/\omega_o)^{1/a}} f'(p',x',\omega J',\phi',\omega),$$

with a = 4, and

(46) $$\| g \|_{\frac{1}{2}\rho} < 2F , \qquad \| f' \|_{\frac{1}{2}\rho} < 2F .$$

In particular, for real initial data, one has

(47) $$\begin{aligned} |E'(t) - E'(0)| < 4\omega^{-1} F/\rho, &\qquad |E(t) - E(0)| < 8\omega^{-1} F/\rho \\ \text{for } |t| &\leq e^{(\omega/\omega_o)^{1/a}} \end{aligned}$$

Remarks: - The value a = 4 is not optimal; a better proof should give

$a = 2 - \eta$, with η as small as one likes.

- The number n of degrees of freedom is irrelevant, as none of the above estimates contains explicitely n. Of course, one could be interested to F proportional to n; in that case the n-dependence turns out to be optimal, as is easily seen on counterexamples (the point is that one must take care of those small regions of the phase space where all of the energy is concentrated in a few degrees of freedom).

- The restriction $J \geq \frac{1}{2}E_o$ in the definition (40) is purely technical; it was introduced to avoid the non-analiticity of the square roots at zero. This is clearly a spurious difficulty, which only reflects the non-analiticity of the transformation (32) to action-angle variables at zero action. To avoid this problem one should work out perturbation theory in the original π, ξ variables: this is possible, but painful (see later for the case $\nu > 1$).

2.3 Let us provide a numerical example of a problem of realization of constraints. Consider a one-dimensional system of n+1 point particles of mass m in a box of length L, interacting via a short-range nearest-neighbours potential $V(r) = V_o \dfrac{e^{-(r/r_o)^2}}{r/r_o}$, with periodic boundary conditions; denote by $x_o, x_1, \ldots, x_n$ and $p_o, p_1, \ldots, p_n$ the coordinates and momenta of the particles. Now introduce the additional potential $\frac{1}{2} m\omega^2 x_o^2$, which for large ω should constrain the zeroth particle to the origin. The full Hamiltonian is then

$$(48) \qquad H = \frac{p_o^2}{2m} + \frac{m\omega^2}{2} x_o^2 + \sum_{i=1}^{n} \frac{p_i^2}{2m} + \sum_{i=0}^{n} V(x_{i+1} - x_i) .$$

In order to define the time scale associated to the energy exchange between the zeroth and the remaining degrees of freedom, let us focus the attention to the vibrational energy $E = \frac{1}{2}(p_o^2/m + m\omega^2 x_o^2)$, and consider its time correlation

$$(49) \qquad G(\tau) = \frac{<E(\tau)E(0)> - <E(\tau)><E(0)>}{<E(0)^2> - <E(0)^2>}$$

where <.> denotes an average on the initial data, with the canonical probability $\propto e^{-\beta H}$ (which is easily realized via a Monte-Carlo procedure).

Preliminary results performed on a model with $n = 4$, $L = 4(n+1)r_o$ using r_o, m, V_o as the units of length, mass and energy ($t_o = r_o m^{\frac{1}{2}} V_o^{-\frac{1}{2}}$ is then the unit of time), indicate that $G(\tau)$ decays quite rapidly with τ, from one to values close to zero, as shown for example in figure 1. The time τ_c needed in order $G(\tau)$ reaches any fixed value, say $G(\tau_c) = \frac{1}{2}$, can be used as an empirical definition of the time scale associated to a sensible energy exchange between the "constrained" degree of freedom and the remaining ones.

As a matter of fact, τ_c increases quite rapidly with ω, as shown in figure 2, which represents log τ_c vs. ω for two different values of β ($\beta = V_o^-$ and $\beta = \frac{1}{2}V_o^-$ for circles and triangles respectively). For both values of β the data are quite well interpolated by straight lines, which correspond to exponential laws of the form

$$(50) \qquad \tau_c(\beta,\omega) = A(\beta)e^{\omega/\omega_o(\beta)} \; ;$$

for $\beta = V_o^{-1}$ one has $A \simeq 0.5$, $\omega_o \simeq 0.8$. By comparison with (47), it can be seen that one has $a = 1$; this is somehow surprising, because it is even better than the optimal value $a = 2$ (i.e., the value of Arnold examples of diffusion [4]). In my opinion, this does not depend on a lucky choice of the model, but on the average introduced in the definition of τ_c: the value $a = 2$ is probably optimal for a small subset of the phase space, which is determinant for the rigorous estimates of perturbation theory, but is not so relevant in the average leading to (50). Let me stress that the exponential law (50), if extrapolated to high frequencies, gives times scales increasing dramatically with ω : for example, if in the above model we assume that the zeroth particle is constrained to the origin by a violin LA-string, of the same mass m as our particles, then one has $\omega = \sqrt{2}\pi\nu \simeq 2 \times 10^3 \sec^{-1}$, and assuming that r_o and V_o are such that $t_o = 1$ sec (say, particles of mass 1 gr, at average energy 1 erg, smoothly interacting on a range of 1 cm), one would find a time scale greater than the universe lifetime in order that collisions alter appreciably the intensity of the violin sound.

Up to here we have only considered the case $\nu = 1$; a few comments on the case $\nu > 1$ will be reported in the next section, after a comment on the possible relevance of the exponential estimates like (50) for statistical mechanics.

3. LONG TIME SCALES IN STATISTICAL MECHANICS

It is a remarkable fact that long time before Arnold's examples of diffusion on exponentially long times, and before Nekhoroshev exponential estimates, Boltzmann and Jeans clearly understood the possible existence of very long time scales in classical dynamical systems, associated to the energy sharing among translational and vibrational degrees of freedom.

The aim of Boltzmann and Jeans was to explain classically some phenomena later considered to be essentially quantum: for example, to explain why in a gas of diatomic molecules the internal degree of freedom does not contribute to the specific heat, or why in a blackbody one does not see the continuous energy flow from the low frequency to the high frequency degrees of freedom, leading to the ultraviolet catastroph, as expected in agreement with the equipartition principle. Boltzmann intuition, expressed in a letter to Nature in 1895

[12], was that the time scale associated to equipartition could be so long ("days or years", in his own words), that in ordinary experiments the high frequency degrees of freedom would behave (using a later terminology) as if they were frozen.

Boltzmann conjecture has apparently no support but his intuition. On the contrary, Jeans [13] [14] took into consideration several model examples, and working heuristically on them he arrived to conceive the existence of time scales of "billions of years", in order the high frequency degrees of freedom acquire a significant amount of energy. Moreover, as in my opinion is quite impressive, an exponential law analogous to (50), i.e. with a = 1, also appears in Jeans papers.

Take, as a model example, a one dimensional system of diatomic molecules, like the one represented in figure 3. Denote by $x_1,\ldots,x_n$ the coordinates of the centers of mass of the molecules, and by $\xi_1,\ldots$ $\ldots,\xi_n$ the internal degrees of freedom, say the coordinates of the black particles relative to their centers of mass; assume that both atoms have equal mass $\frac{1}{2}m$, and that springs have vanishing proper length (atoms can cross, and oscillate around a common equilibrium position). Suppose that black particles interact via a short-range nearest-neighbours repulsive potential $V(r) = V_\circ \dfrac{e^{-(r/r_\circ)^2}}{r/r_\circ}$, and that one has fixed-ends boundary conditions, i.e., $x_\circ = 0$, $x_{n+1} = L$, $\xi_\circ = \xi_{n+1} = 0$. The Hamiltonian is then

$$(51)\qquad H = \frac{1}{2}\sum_{i=1}^{n}(\pi_i^2/m + m\,\omega^2\xi_i^2) + \frac{1}{2}\sum_{i=1}^{n} p_i^2/m + \sum_{i=1}^{n+1} V(x_i+\xi_i-x_{i-1}-\xi_{i-1}),$$

$\pi_1,\ldots,\pi_n,p_1,\ldots,p_n$ denoting here the conjugate momenta to $\xi_1,\ldots,\xi_n$, $p_1,\ldots,p_n$. For large ω this gives nothing but the realization of the constraint $\xi_1 = 0,\ldots,\xi_n = 0$, i.e., the constraint of rigidity of molecules. Thus, what we need is a generalization of the considerations of the previous section to $\nu > 1$.

Perturbation theory becomes quite complicated, although not seriously difficult. Making reference to the above example, it is quite clear that the equality of frequencies produces quite rapid energy exchanges of the vibrational energy from spring to spring during collisions: however, this is not a problem, as one only needs to control, for long times, the overall vibrational energy $E = \frac{1}{2}\sum_{i=1}^{n}(\pi_i^2/m+m\omega^2\xi_i^2)$ (one must use the so-called resonant construction of perturbation theory: having here exactly n-1 independent resonance relations, one can construct at most one integral of motion; a nice consequence is that small denominators never appear). There is only a technical difficulty indeed, we cannot use, as for $\nu = 1$, the action-angle variables, because actions are no more bounded away from zero. The estimates become then painfull, but can be done; the conjectured result (in fact, practically proven apart from a few details) is that for systems like the above one has, as in the case $\nu = 1$,

$$(52) \qquad |E(t) - E(0)| \leq \omega^{-1}K \qquad \text{for } |t| \leq e^{(\omega/\omega_o)} \quad ,$$

where a = 4, while K and ω_o depend on the analiticity of the Hamiltonian and on the norms, but not on the number n of degrees of freedom.

Some numerical results on the above model are already available [15]. In place of the phase correlation of E, the time correlation has been here considered in the definition of $G(\tau)$, precisely

$$(53) \qquad G(\tau) = \frac{\overline{E(t+\tau)E(t)} - \overline{E(t)}^2}{\overline{E(t)^2} - \overline{E(t)}^2} \; ,$$

where the overbar denotes the average over the dummy variable t, up to a large time T_{max} which, for consistency, must be much larger than the time over which G is sensibly different from zero; for more details, see [15].

As in the case $\nu = 1$, here too one defines τ_c by the requirement $G(\tau_c) = \frac{1}{2}$; τ_c becomes then a function of ω, of the energy per particle $U = E/n$, and possibly of n. The results obtained up to now are summarized in figure 4, which refers to n = 16 (open symbols), or n = 64 (full symbols); circles and triangles refer respectively to $U = V_o$ and $U = 2V_o$. The length of the box containing the molecules was $4(n+1)r_o$; m, r_o and V_o have been used ad units of measure.

As the figure shows, one finds for τ an exponential law of the form

$$(54) \qquad \tau_c(\omega,U) = A(U)e^{\omega/\omega_o(U)} ,$$

with apparently no n dependence. For $U = V_o$ one has $A \simeq 2t_o$ and $\omega_o \simeq 2t^{-1}$, where $t_o = r_o m^{\frac{1}{2}} V_o^{-\frac{1}{2}}$ is the natural unit of time.

The model is not sufficiently realistic to have any physical significance. Nevertheless, it is worthwhile to remark that the above exponential law can lead to the macroscopic times invoked by Boltzmann and Jeans, if one takes for the units r_o, m and V_o, as well as for ω, suitable microscopic values: for example, for $r_o = 10^{-7}$ cm, m = 2 atomic mass units, $V_o = 0.1$ eV , then one has $A \simeq 2.8 \times 10^{-13}$ sec and $\omega_o \simeq 1.4 \times 10^{13}$ sec^{-1}, while the specific energy $U = V_o$ corresponds to a temperature of about 700°K. One then finds $\tau_c = 1$ sec at $\omega = \omega \simeq 4.1 \times 10^{14}$ sec^{-1}, while at $\omega = 2\,\omega_1$ one finds $\tau_c > 10^5$ years. Of course, this requires a rather bold extrapolation of the data of figure 4 (precisely, to values of ω about ten times larger than the highest value there appearing).

The Boltzmann and Jeans conjecture needs clearly further deeper investigation; in particular, one should also consider systems with more thant one frequency, and systems containing rotational beside translational and vibrational degrees of freedom. Anyhow, the possible connection of such an old conjecture with the most recent result of classical perturbation theory, via the problem of the realization of constraints, is in my opinion quite stimulating, and I warmly hope that working in this direction can lead to a deeper understanding of the relations between classical and quantum statistical mechanics.

REFERENCES

[1] Giorgilli, A. and Galgani, L.: "Celestial Mechanics" 37, 95(1985)

[2] Poincaré, H.: "Les Méthodes Nouvelles de la Méchanique Céleste", Vol. 3 (Gautier-Villars, Paris, 1899)

[3] Nekhoroshev, N.N.: "Usp. Mat. Nauk 32, 5 (1977) Russ. Math. Surv. 32, 1 (1977) ; Trudy Sem. Petrows. No. 5, 5 (1979)

[4] Arnold, V.I.: Dokl. Akad. Nauk. SSSR 156, 9 (1964) Sov. Math. Dokl. 6, 581 (1964)

[5] Gallavotti, G.: lectures given at the 1984 Les Houches Summer School, to be published

[6] Benettin, G. and Gallavotti, G.:"Exponential Estimates for the Stability times in Nearly Integrable Hamiltonian Systems", to appear in J. Stat. Phys.

[7] Benettin, G., Galgani, L., Giorgilli, A.: "Celestial Mechanics 37, 1 (1985)

[8] Gallavotti, G.: "Meccanica Elementare (Boringhieri, Torino, 1980) English edition: "The Elements of Mechanics (Springer Verlag, Berlin, 1983)

[9] Rubin, H., HUngar, G.: Commun. Pure and Appl. Math. 10, 65 (1957)

[10] Tackens, F.: "Motion under the Influence of a Strong Constraining Force", in Global Theory of Dynamical Systems, L.N.M. 819, Z. Nitecki and C. Robinson editors (Springer Verlag, Berlin 1979)

[11] Benettin, G., Galgani, L., Giorgilli, A.: "Classical Perturbation Theory and the Realization of the Holonomic Constraints", paper in preparation

[12] Boltzmann, L.: "Nature" **51**, 413 (1985)

[13] Jeans, J.H.: Phil. Mag. 6, 279 (1903)

[14] Jeans, J.H.: Phil. Mag. 10, 91 (1905

[15] Benettin, G., Galgani, A., Giorgilli, A.: "Exponential Law for Equipartition Times among Translational and Vibrational Degrees of Freedom", preprint 1986.

REMARKS ON SEIFERT'S PROBLEM AND NONLINEAR COMPLEX EIGENVALUES

Henri Berestycki* and Jean-Michel Lasry**

*Université Paris Nord, Dept. de Mathématiques
93430 - Villetaneuse France

**Ceremade, Université Paris-Dauphine
Place du Maréchal de Lattre
75775 Paris cedex 16 - France

The existence of periodic orbits for some systems of differential equations is considered here. In particular, we are concerned with the existence of periodic orbits for some conservative systems. That is, a system

$$(0.1) \qquad \dot{x} = \phi(x)$$

where $\phi \in C(\mathbb{R}^{2N}, \mathbb{R}^{2N})$ such that along the trajectories $x(t)$, some quantity $G(x(t))$ is conserved, with $G \in C^1(\mathbb{R}^{2N}, \mathbb{R})$. In other words, we assume that the field ϕ satisfies a condition of the form:

$$(0.2) \qquad \phi(x) \cdot G'(x) = 0 \ , \quad \forall\, x \in \mathbb{R}^{2N}$$

($G'(x)$ denotes the gradient of G at x).

The approach to the question that we develop here consists in studying an associated nonlinear complex eigenvalue problem. The method we use combines an a priori estimate technique with a new topological method for the existence of solutions. A particular use is made of the S^1-invariance of the problems which are considered. We report here on results from our work [2] to which the reader is referred for the general proofs and a more complete presentation.

1. Seifert's problem

The model case of a conservative system is when $G(x) = \frac{1}{2}|x|^2$. That is, the field $\phi(x)$ is tangential to spheres:

$$(1.1) \qquad \phi(x) \cdot x = 0 \qquad \forall\, x \in \mathbb{R}^{2N}$$

Then, the trajectories of the system (0.1) lie on spheres

$$S_R = \{ x \in \mathbb{R}^{2N},\ |x| = R \}$$

(Notice that if the dimension were odd, that is, with $\phi \in C(\mathbb{R}^{2N+1}, \mathbb{R}^{2N+1})$, by the Hedgehog theorem, there is always a rest point for a tangential vector field, whence a trivial periodic orbit. This is why we only consider systems in even dimensions).

It was originally conjectured that such a system always has at least one periodic orbit on a given sphere. But owing to some counter-examples, we now know that such a general result is not true. It is not yet clear, however, whether or not this conjecture holds for smooth mappings. That is, whether or not any C^∞ (say) mapping ϕ satisfying $\phi(x) \cdot x = 0$, $\forall\, x \in S_R$ admits at least one periodic orbit on the sphere S_R.

Here, we are concerned with a different modification of this conjecture which is more motivated by the theory of Hamiltonian systems. Consider the standard skewsymmetric matrix in $\mathbb{R}^{2N}$:

$$I = \begin{pmatrix} 0 & -I_N \\ I_N & 0 \end{pmatrix}$$

(I_N stands for the identity in $\mathbb{R}^N$). One imposes the supplementary condition:

(1.2) $\phi(x) \cdot Jx > 0, \quad \forall\, x \in S_R$.

Conjecture: Let $\phi : \mathbb{R}^{2N} \to \mathbb{R}^{2N}$ be a continuous mapping satisfying (1.1) (on S_R) and (1.2). Then, there exists a periodic orbit of (0.1) on S_R.

This problem is still open. Local results with R very small (see next section) as well as small perturbation results have been proved. For the latter, assuming that $\phi(x)$ is a sufficiently small perturbation of the field Jx, existence of periodic orbits was shown by Seifert [14], Fuller [7] and Reeb. Here we establish a somewhat more precise and more global result in the direction of this conjecture.

Theorem 1. Let $\phi \in C^\circ(S, \mathbb{R}^{2N})$ be a mapping such that $\phi(x) \cdot x = 0$ and $|\phi(x) - Jx| < \frac{1}{3}|x|$, $\forall\, x \in S_R$. Then, there exists at least one periodic orbit of (0.1) on S_R.

Similar results for more general conservative systems will be stated in section 3. The preceding Theorem will be seen in section 4 to be a straightforward consequence of the next result concerning systems which are not necessarily conservative.

Theorem 2. Let $\phi \in C^\circ(\mathbb{R}^{2N}, \mathbb{R}^{2N})$ satisfy $|\phi(x) - Jx| \leq \alpha|x|$, $\forall\, x \in \mathbb{R}^{2N}$ with $0 < \alpha < 1/3$. Then, for any $R > 0$, there exist real numbers λ, μ

with $1/2 \leq \lambda \leq 3/2$ and a 2π-periodic solution $x(t)$ for the system

$$(1.3) \qquad \dot{x} = (\lambda I + \mu J)\,\phi(x)$$

such that $\| x \|_{L^2(0,\pi)} = R$

(Here, I stands for the identity in $\mathbb{R}^{2N}$). The proof of this Theorem will be divided into sections 5 to 7. It will be seen that (1.3) reduces to a nonlinear complex eigenvalue problem, where the unknown "eigenvalue" is $\zeta = \lambda + i\mu$. (See section 5)

Lastly, the topological method we use (section 7) allows us to derive some results in section 8 for nonlinear complex eigenvalues in an abstract setting. General results for this kind of problems have also been obtained by J. Ize [9].

2. Conservative Systems

The notion of conservative system, i.e. a system (0.1) with ϕ satisfying (0.2), is broader than that of Hamiltonian system. Indeed, let $H \in C^1(\mathbb{R}^{2N}, \mathbb{R})$ and consider the system of 2N equations $(1 \leq i \leq N)$:

$$\dot{p}_i = -\frac{\partial H}{\partial q_i}(p,q) \qquad\qquad \dot{q}_i = \frac{\partial H}{\partial p_i}(p,q)$$

or, writing $x = (p,q) \in \mathbb{R}^{2N}$:

$$(2.1) \qquad \dot{x} = JH'(x).$$

This is a conservative system: the "conserved quantity" here simply is the Hamiltonian $H(x)$.

This notion arises naturally in the study of periodic orbits near a stationary point. Let $\phi \in C^1(\mathbb{R}^{2N}, \mathbb{R}^{2N})$ be such that $\phi(0) = 0$ and

$\phi'(0)$ is invertible. One is interested in the existence of periodic orbits of the system $\dot{x} = \phi(x)$ which are arbitrarily close to 0. A necessary condition for this is that $\phi'(0)$ possesses a pair of eigenvalues $\pm i\omega$ with $\omega \in \mathbb{R}\setminus\{0\}$. This condition, however, is not sufficient in general as is readily seen from the following example in $\mathbb{R}^2$:

$$(2.2) \qquad \left.\begin{aligned} \dot{x}_1 &= -x_2 - x_1(x_1^2 + x_2^2) \\ \dot{x}_2 &= x_1 - x_2(x_1^2 + x_2^2) \end{aligned}\right\} = \phi(x_1, x_2)$$

Indeed, in this case, $\phi'(0) = \begin{pmatrix} 0 & -1 \\ 1 & 0 \end{pmatrix}$ and solutions of (2.2) satisfy $r\dot{r} = -r^4$ where $r^2 = x_1^2 + x_2^2$. Hence (2.2) has no periodic solution apart from the constant solution $x_1 = x_2 = 0$. Therefore, one is led to impose other conditions.

In particular, for a large class of conservative systems the preceding condition is actually sufficient to ensure the existence of periodic orbits close to 0. The first result in this direction is the celebrated "center Theorem" of Liapunov [11]. A more general and simpler version of this result (avoiding "non degeneracy" assumptions) has been given by Moser 12 who established the following

<u>Theorem [12].</u> Let $\phi \in C^1(\mathbb{R}^{2N}, \mathbb{R}^{2N})$, $G \in C^1(\mathbb{R}^{2N}, \mathbb{R})$ satisfying $\phi(x) \cdot G'(x) = 0$, $\forall\, x \in \mathbb{R}^{2N}$. Assume furthermore that $G(0) = 0$, $G'(0) = 0$ and that $G''(0)$ is positive definite. Then, for all $\varepsilon > 0$ sufficiently small there exists at least one periodic orbit of (0.1) on the surface $\{G = \varepsilon\}$.

It is further shown in [12] that a multiplicity result obtains under certain nondegeneracy assumptions and that the periods of the solutions are close to the ones of the linearized system $\dot{x} = \phi'(0)x$.

In the next section, we will state a more global and general result for conservative systems.

For the particularly important class of Hamiltonian systems, more global results are known as well as multiplicity results. We refer to the works of Weinstein [15] , Rabinowitz [13], Ekeland & Lasry [6], Berestycki, Lasry, Mancini & Ruf [3] (see also the survey in [1] and the bibliography therein). We also refer to the recent works on Ekeland [4] and Ekeland & Hofer [5]. In particular we recall the result of Rabinowitz [13]. If $\{H = R\}$ is a manifold and bounds a starshaped region, then the Hamiltonian system (2.1) has at least one periodic orbit on this surface. Ekeland and Hofer [5] have shown that this system has at least two periodic orbits if $\{H = R\}$ bounds a convex region.

The main difference between general conservative systems and Hamiltonian systems is the lack of variational formulation ("Principle of least action") for the former. Actually there is a sharp contrast between the two classes regarding multiplicity of periodic orbits. For Hamiltonian systems one hopes, in general, to find "many" periodic solutions (Cp. [6], [3], [4] and [5]). Whereas an example due to Moser [12] shows that for conservative system, this is not the case. The system is in $\mathbb{R}^4$ and, with the notations $w_1 = x_1 + i x_3$ $w_2 = x +$ $+ i x_4$, it reads:

$$\dot{w}_1 = i k w_1 + \bar{w}_2 w_2^{k+1}$$

$$\dot{w}_2 = i w_2 - \bar{w}_1 w_2^{k+1}$$

with k an integer. This system satisfies condition (1.1) (trajectories are on spheres). And on each sphere there is only one periodic orbit (See [12]).

The question of knowing whether or not conservative systems do have generically (in a sense e.g. analogous to that given in [4]) many solutions on energy surfaces $\{G = \text{constant}\}$ remains open.

3. Existence of periodic orbits for conservative systems

We now state our result in a general setting. Let S be a $2N \times 2N$ skewsymmetric invertible matrix. We denote by $\pm i\omega_1, \ldots, \pm i\omega_N$ its eigen values ($\omega_1, \ldots, \omega_N > 0$). For each k $\mathbb{N}$, $1 \leq k \leq N$, we associate to ($\omega_1, \ldots, \omega_N$) constants α_k and δ_k defined in the following manner. For each $\rho \in (0,1)$, let

$$(3.1) \qquad \mu_m(\rho) = \underset{j \in \mathbb{Z}}{\text{Min}} \left\{ \omega_k \frac{\left| j - \frac{\omega_\ell}{\omega_k}(1 \pm \rho) \right|}{(1 \pm \rho)} \right\}$$

$$\ell = 1,2,\ldots,N$$

Then, for all $\rho \in (0,1)$ such that $\frac{\omega_\ell}{\omega_k}(1 \pm \rho) \notin \mathbb{Z}$ for each $\ell = 1, 2, \ldots, N$, the number $\mu_k(\rho)$ is positive. We define

$$(3.2) \qquad \alpha_k = \underset{0 < \rho < 1}{\text{Max}} \ \mu_k(\rho)$$

It is easily seen that $\exists \rho_k \in (0,1)$ such that

$$(3.3) \qquad \alpha_k = \mu_k(\rho_k)$$

Finally, we set

$$(3.4) \qquad \delta_k = 2\pi \rho_k$$

We assume in this section that $G \in C^1(\mathbb{R}^{2N}, \mathbb{R})$ satisfies the following conditions

(3.5) $\begin{cases} \text{The surface } \Sigma = \{G = R\} \text{ bounds a bounded open set } \Omega = \{G<R\} \\ \text{which is strictly starshaped with respect to } 0. \end{cases}$

(3.6) $G' \neq 0$ on Σ

We further assume that $\phi \in C^\circ(\Sigma, \mathbb{R}^{2N})$ satisfies:

(3.7) $\phi(x) \cdot G'(x) = 0 \ , \ \forall \ x \in \Sigma,$

(3.8) $S\phi(x) \cdot G'(x) < 0 \ , \ \forall \ x \in \Sigma$

Our main result is the following

Theorem 3. Let $\phi \in C^\circ(\Sigma, \mathbb{R}^{2N})$ and $G \in C^1(\mathbb{R}^{2N}, \mathbb{R})$ satisfy conditions (3.5) - (3.8). For all $k \in \mathbb{N}$, $1 \leq k \leq N$, let α_k and δ_k be defined as above. If ϕ further satisfies

(3.9) $|\phi(x) - Sx| < \alpha_k \ |x| \ , \ \forall \ x \in \Sigma$

then, there exists at least one periodic orbit of (0,1) on $\Sigma = \{G = R\}$. Moreover, this solution has a period T which satisfies $|T - \frac{2\pi}{\omega_k}| < \delta_k$.

Remark. Theorem 1 is a particular case of the preceding one, corresponding to the case $S = J$ and $G(x) = \frac{1}{2} |x|^2$. Indeed when $S = J$, i.e. $\omega_1 = \omega_2 = \ldots = \omega_N = 1$, one has, for all $k = 1,2,\ldots,N$,

$$\mu_k(\rho) = \mu(\rho) = \text{Min} \ \{\frac{\rho}{1+\rho} \ ; \ \frac{1-\rho}{1+\rho}\}$$

One then obviously has $\rho_k = \rho = 1/2$ and

$$\alpha_k = 1/3 , \quad \delta_k = \pi, \qquad \forall\ k = 1,2,\ldots,N.$$

For the sake of simplicity, in the following we restrict ourselves to this case ($S = J$ and $G(x) = \frac{1}{2}|x|^2$). And we start with the proof of Theorem 1.

4. Derivation of Theorem 1

We indicate here how Theorem 1 is a straightforward consequence of Theorem 2 (see section 1). Indeed, let ϕ be as in Theorem 1. We may extend it to be defined as a continuous mapping $\phi : \mathbb{R}^{2N} \to \mathbb{R}^{2N}$ such that $|\Phi(x) - Jx| \leq \alpha|x|$, $\forall\ x \in \mathbb{R}^{2N}$ with $0 \leq \alpha \leq 1/3$. Let $x(t)$ be the 2π-periodic solution of (1.3) given by Theorem 2 and such that

$$\| x \|_{L^2(0,\pi)} = R/\sqrt{2\pi}.$$

Setting $y(t) = x(t/\lambda)$ we obtain a periodic (with period $2\pi\lambda$) solution y of the system:

$$\dot{y} = \phi(y) + \nu J \phi(y) \tag{4.1}$$

for the real $\nu = \mu/\lambda$. Multiplying (4.1) by y and integrating over $(0,T)$ ($T = 2\pi\lambda$ is the period) we get, using condition (1.1):

$$0 = \int_0^T \dot{y} \cdot y\, dt = \nu \int_0^T J\phi(y) \cdot y\, dt . \tag{4.2}$$

As a consequence of $|\phi(x) - Jx| \leq \alpha|x|$, we know that $J\phi(x) \cdot x < 0$ $\forall\ x \neq 0$. Since y is not trivial (i.e. $y \not\equiv 0$), (4.2) implies that $\nu = 0$. Hence, the function $y(t)$ is actually a periodic solution of (0.1).

Then, by (1.1), $|y(t)|$ is constant, whence also $|x(t)|$ is constant. From the condition $\| x \|_{L^2(0,\pi)} = R/\sqrt{2\pi}$ we then derive that $y(t) \in S_R$ $\forall$ t. The derivation of Theorem 1 is thereby complete.

5. A nonlinear complex eigenvalue problem

In this and the next two sections, we prove Theorem 2. We identify $\mathbb{R}^{2N}$ with $\mathbb{C}^N$ via the isomorphism $(x_1,\dots,x_N, y_1,\dots,y_N) \longleftrightarrow (x_1 + i y_1,\dots,x_N + i y_N)$. Multiplication by the matrix I in $\mathbb{R}^{2N}$ then corresponds to scalar multiplication by i in $\mathbb{C}^N$. Henceforth, we consider ϕ as a continuous mapping: $\mathbb{C}^N \to \mathbb{C}^N$. Thinking of 2π-periodic functions as being defined on S we seek a solution as a mapping x: $S^1 \to \mathbb{C}^N$. We let $E = L^2(S^1,\mathbb{C}^N)$ with the usual scalar product. Thus, the problem we want to solve in Theorem 2 (section 1) now reads:

Problem 5.1. To find $x \in E$ and $\zeta = \lambda + i\mu \in \mathbb{C}$ satisfying $\dot{x} = \zeta \phi(x)$, $| x |_E = R$ and $1/2 < \operatorname{Re} \zeta < 3/2$.

We now face a nonlinear complex eigenvalue problem.

For technical reasons, we require a finite dimensional reduction. We denote by E^m the space of Fourier series truncated at order m:

$$E^m = \{x \in E;\quad x = \sum_{-m}^{m} x_k e^{ikt}, \quad x_k \in \mathbb{C}^N\}.$$

Denote by Q_m the orthogonal projector from E onto E^m. The finite dimensional approximation of Problem 5.1 is the following:

Problem 5.2. To find $x \in E^m$ and $\zeta \in \mathbb{C}$ with $\dot{x} = \zeta Q_m[\phi(x)]$, $\| x \|_E = R$, $1/2 < \operatorname{Re} \zeta < 3/2$.

Let us indicate how one obtains a solution of Problem 5.1 by passing to the limit in Problem 5.2. Let (x_m,ζ_m) be a solution of the latter for each $m\in\mathbb{N}^*$. By the assumption on ϕ, $\phi(x_m)$ is clearly bounded in L^2. Hence $Q_m[\phi(x_m)]$ is also bounded in L^2. Provided we have a uniform bound on $|\zeta_m|$ (in fact, it will be shown here that $|\zeta_m -1| \leq 1/2$) we know that $\|\dot{x}_m\|_{L^2}$ is bounded, that is x_m is bounded in $H^1(S^1,\mathbb{C}^N)$. Since $H^1(S^1,\mathbb{C}^N)\hookrightarrow L^\infty(S^1,\mathbb{C}^N)$ with compact injection. We may strike out a subsequence x_{mj}, ζ_{mj} such that $x_{mj} \to x$ uniformly and $\zeta_{mj} \to \zeta$. Therefore $Q_{mj}[\phi(x_{mj})]$ converges strongly in L^2 to $\phi(x)$. This shows that (x,ζ) is a solution of Problem 5.1.

6. An a priori estimate technique

In order to prove existence of solutions for problem 5.1 or 5.2 we require some a priori estimates. Since a homotopy argument will be used, we require such an estimate for a family of equations which we define now. Let $s\in[0,1]$ and define

(6.1) $$\phi^s(x) = (1-s)\,\phi(x) + i\,s\,x$$

(6.2) $$H^s_\zeta(x) = i\,\zeta^{-1}\,\dot{x} - i\,Q_m[\phi^s(x)].$$

The result we require is the following.

Lemma 6.1. For all $s\in[0,1]$ and for all ζ on the circle $\{\zeta\in\mathbb{C},\ |\zeta-1| = 1/2\}$, $H^s_\zeta(x) \neq 0$ for all $x\in E^m\setminus\{0\}$.

Proof. By the assumption on ϕ we also have

(6.3) $$|\phi^s(x) - i\,x| \leq \alpha|x| \qquad \forall\, x\in\mathbb{C}^N$$

with $0 < \alpha < 1/3$. Suppose that there exists a solution $x \neq 0$, $x\in E^m$, of the equation $H^s_\zeta(x) = 0$. That is,

(6.4) $\dot{x} = \zeta\, Q_m[\phi^s(x)]$.

Hence,

(6.5) $\| \dot{x} - \zeta i\, x \|_{L^2} = |\zeta|\ \| Q_m[\phi^s(x) - ix] \|_{L^2}$

By (6.3) and since Q_m is a projection, we derive:

(6.6) $\| \dot{x} - \zeta\, i\, x \|_{L^2} \leq \alpha|\zeta|\ \| x \|_{L^2}$.

Write $x = \sum_{-m}^{m} a_k\, e^{ikt}$ and (6.6) reads

(6.7) $\sum_{-m}^{+m} |k - \zeta|^2\ |a_k|^2 \leq \alpha^2|\zeta|^2 (\sum_{-m}^{m} |a_k|^2)$.

Since $|\zeta - 1| = \frac{1}{2}$, we have $\alpha^2\,|\zeta|^2 < \frac{1}{4}$ while $|k - \zeta|^2 \geq \frac{1}{4}$ for all $k \in \mathbb{Z}$. This obviously contradicts (6.7) if $x \neq 0$. The proof of Lemma 6.1 is thereby complete.

7. A topological method for the existence of solutions

We now prove the following

Proposition 7.1. For each $m \in \mathbb{N}^*$, there exists a solution (x, ζ) of Problem 5.2 with $|\zeta - 1| \leq 1/2$.

On the spaces E or E^m, S^1 acts through time shifts. Indeed, for $\tau \in \mathbb{R}/2\pi\mathbb{Z}$, we let

$$T_\tau x(t) = x(t + \tau) .$$

This action leaves the spaces E and E^m invariant. Notice that this action is not free as it admits the space E° of fixed points. We denote by S the sphere

$$S = \{x \quad E^m ; \ \| x \|_E = R \} .$$

A mapping $\Psi : S \to S$ is termed <u>equivariant</u> if $\Psi \circ T_\tau = T_\tau \circ \Psi$, $\forall \tau \in \mathbb{R}/2\pi\mathbb{Z}$. The space of all continuous and equivariant mappings: $S \to S$ will be denoted by X.

<u>Proof of Proposition 7</u>. We argue by contradiction and assume that for some m:

$$(7.1) \qquad \begin{cases} \dot{x} - \zeta Q_m[\phi(x)] \neq 0 \\ \forall x \in S , \quad \forall \zeta \in \mathbb{C} , \quad |\zeta - 1| \leq 1/2 \end{cases}$$

For all $\zeta \in \mathbb{C}$, $|\zeta - 1| \leq 1/2$; one can then define a mapping $\Psi_\zeta \in C(S,S)$ by setting

$$\Psi_\zeta(x) = R \frac{i \zeta^{-1} \dot{x} - i Q_m [\phi(x)]}{\| i \zeta^{-1} \dot{x} - i Q_m [\phi(x)] \|_{L^2}} .$$

It is easily seen that $\Psi_\zeta \in X$.

We now define:

$$\gamma_\rho(\theta) = \Psi_{1 + \rho e^{i\theta}} , \quad 0 \leq \rho \leq 1/2$$

For each fixed $\rho \in [0,1/2]$, γ_ρ defines a loop in X, that is, γ_ρ is an element of $\pi_1(X)$ (the base point being the identity). When ρ goes from 1/2 to 0, our starting assumption (7.1) shows that $\gamma_{1/2}$ and γ_0 are homotopic, whence since γ_0 is constant:

(7.2) $\gamma_{1/2} \simeq 0$

But we have constructed in section 6 (see (6.1) and (6.2)) another deformation H_ζ^S. More precisely, we define

$$\varphi_\zeta^S(x) = R \frac{H_\zeta^S(x)}{\| H_\zeta^S(x) \|_{L^2}}$$

Clearly, $\varphi_\zeta^S \in X$ and we let

$$g^S(\theta) = \varphi^S_{1 + (e^{i\theta}/2)}$$

Again g^S defines an element of $\pi^1(X)$ and letting s vary in [0,1] we have shown that g^1 and g° are homotopic. Since $\varphi_\zeta^\circ = \Psi_\zeta$, we have $g^\circ = \gamma_{1/2}$ and therefore, in view of (7.2) we know that

(7.3) $g^1 \simeq 0$ in $\pi^1(X)$.

Notice that, up to a constant factor,

$$g^1(\theta)[x] = x + i\,\zeta^{-1}\,\dot{x}$$

Letting $x = \sum_{-m}^{m} x_k\, e^{ikt}$, this means:

$$g^1(\theta)[x] = \sum_{-m}^{+m} (1 - k\zeta^{-1})\, x_k\, e^{ikt}$$

Now, for all $k \neq 1$, and for all $s \in [0,1]$, for $\zeta \in \mathbb{C}$ with $|\zeta - 1| = 1/2$, we have

$$1 - k(s\zeta + 1 - s)^{-1} \neq 0 .$$

It follows that g^1 is homotopic to the loop $\hat{g}(\theta)[x] = e^{i\theta} x_1 e^{it} + \sum_{\substack{k \neq 1 \\ |k| \leq m}} (1 - k)\, x_k\, e^{ikt}$ whence, also to the loop:

(7.4) $$h^1(\theta)[x] = e^{i\theta} x_1 e^{it} + \sum_{\substack{k \neq 1 \\ |k| \leq m}} x_k e^{ikt}$$

Thus, the assumption (7.1) leads to

(7.5) $$h^1 \simeq 0 \quad \text{in } \pi_1(X)$$

One can further specify the way all the preceding deformations act on E°, the space of fixed points of the action. Firstly it is readily checked that they all leave the space E° invariant. Consider the following subspaces of X

$$\hat{X} = \{h \in X \; ; \; h(S \cap E^\circ) \subset E^\circ\}$$

$$X_1 = \{h \in \hat{X} \; ; \; |h(x) - x| < |x| \, , \; \forall\, x \in S \cap E^\circ\}$$

$$Y = \{h \in \hat{X} \; ; \; h(x) = x \, , \; \forall\, x \in E^\circ\}$$

By inspection of all the deformation which we have used, we obtain, more precisely than in (7.5), that

(7.6) $$h^1 \simeq 0 \quad \text{in } \pi_1(X_1)$$

And since $h^1 \in Y$, it is straightforward to check that (7.6) implies

(7.7) $$h^1 \simeq 0 \quad \text{in } \pi_1(Y).$$

Observe that $h^1(\theta)$ is a unitary linear operator. Denote by Z the space of such operators whose trace on E° is the identity. In particular, $h^1 \in \pi^1(Z)$. In order to conclude the proof, the following result of Husseini [8] is crucial.

Theorem [8]. The natural injection $Z \to Y$ induces a homomorphism $\pi_1(Z) \to \pi_1(Y)$ which itself is injective.

This is a particular case of more general results in this spirit developped by Husseini in [8].

By this theorem then, (7.7) implies the following stronger assertion:

(7.8) $\qquad h^1 \simeq 0 \quad \text{in } \pi_1(Z)$.

This, however, is impossible. Indeed, (7.8) would imply that $\chi(\theta) = \det(h^1(\theta))$ is a trivial loop in $\pi_1(S^1)$, while $\chi(\theta) = e^{in\theta}$.

The proofs of Proposition 7.1 and consequently of Theorem 1 and 2 are thereby complete.

8. Nonlinear complex eigenvalues

Several general results concerning nonlinear complex eigenvalue problems have been established by Ize in [9] and [10]. The methods we have presented here also allow us to prove some existence results in an abstract setting. We also mention here the simplest case in a finite dimensional space. Analogous more general results in infinite dimensional spaces are proved in [2].

Theorem 3. Let T be a free unitary representation of S^1 in $\mathbb{C}^N$ and let $A : \mathbb{C}^N \to \mathbb{C}^N$ be a continuous mapping which is equivariant ($A \circ T_\tau =$

A NOTE ON A THEOREM OF CONLEY AND ZEHNDER

Maria Letizia Bertotti*

Institut für Mathematik - Ruhr Universität Bochum
4630 Bochum (BRD)
Present address: Dipartimento di Matematica Università di Trento
38050 Povo - Trento (I)

ABSTRACT

We give a result concerning existence of two forced oscillations for (periodically) time dependent, asymptotically linear Hamiltonian Systems. The result can be viewed as a kind of generalization to higher dimensions of the Poincaré - Birkhoff fixed point theorem.

1. INTRODUCTION AND RESULT

The result we present here strenghtens a remarkable theorem of Conley and Zehnder, concerning forced oscillations of asymptotically linear Hamiltonian Systems, which can be viewed as a kind of generalization to higher dimensions of the Poincaré-Birkhoff fixed point theorem.

The P.-B. theorem, we recall, states that every measure preserving homeomorphism of an anulus A in the plane R , twisting the two boundaries in opposite directions, has at least two fixed points in the interior of A.

It is a global statement in nature and it merely postulates some qualitative behaviour of the map at the boundary of A and not in its interior, except for the measure preserving character.

So far no genuine higher dimensional analogue of it has been found.

The problem we study is now the following:
we study the time dependent Hamiltonian System

$$(1.1) \qquad \dot{x} = J\nabla h(t,x) \qquad (t,x) \in R \times R^{2n},$$

where the Hamiltonian $h(t,x) \in C^2(R \times R^{2n})$, $n \geq 2$, is periodic in time of period T:

$$h(t+T,x) = h(t,x)$$

We are interested in finding T-periodic solutions of (1.1).

We postulate for this system a certain qualitative behaviour for $|x| \to \infty$ (which can be interpreted as a qualitative behaviour at an "outer boundary") and near the origin, which we assume to be an equilibrium point (and which can be seen as an "inner boundary").

In fact we assume the system to be asymptotically linear and we require for the linearized system at ∞ a nondegeneracy condition (namely we ask its Floquet multipliers to be all $\neq 1$). Now: for this, as for every nondegenerate linear system, it is possible to define a "winding number" ($\in Z$; see [5] for its definition), which we denote in this case by j_∞. This is a Maslov-type index and, roughly speaking, it carries some information on the amount of rotation of the vectorfield at ∞.

Moreover we suppose that also the linearized system at the origin is nondegenerate, and we denote by j_0 its winding number.

The analogy of this situation with P.-B. theorem lies in the fact that the forced oscillations, we are looking for, correspond to the fixed points of the T-map of the flow of (1.1), which is a symplectic map of R^{2n}.

We will not need (in order to deduce existence of forced oscillations) particular estimates nor restrictions on the Hamiltonian vectorfield, apart from the asymptotic behaviour and the requirement that the winding numbers $j_\circ$ and j_∞ are sufficiently different (which corresponds to the twisting of the two boundaries of the annulus in opposite directions).

In [5] the following result has been proved:

Theorem (Conley and Zehnder). Let $h(t,x) \in C^2(R \times R^{2n})$, $n \geq 2$, be periodic in time:

$$h(t+T,x) = h(t,x) .$$

Assume:

i) the hessian of h is bounded:

$$-\beta \leq h_{xx}(t,x) \leq \beta \qquad (t,x) \in R \times X^{2n}, \quad \beta > 0;$$

ii) $J \nabla h(t,x) = JA_\infty(t)x + O(x)$ as $|x| \to \infty$

$J \nabla h(t,x) = JA_\circ(t)x + \dot{O}(x)$ as $|x| \to 0$

uniformly in t for two continuous lops $A_\infty(t+T) = A_\infty(t)$ and $A_\circ(t+T) = A_\circ(t)$;

iii) the two linear systems

$$\dot{y} = JA_\infty(t)y$$

$$\dot{y} = JA_\circ(t)y$$

are nondegenerate. Denote by j_∞ and $j_\circ$ the indices of these two linear systems.

Then, if $j_\circ \neq j_\infty$, there exists a nontrivial T-periodic solution of

$$\dot{x} = J \nabla h(t,x) .$$

Moreover, if this periodic solution is nondegenerate, there is a second nontrivial T-periodic solution.

What we add at this point is the following:

Theorem. Assume the hypotheses in the Theorem above to be satisfied. If $j_\circ \neq j_\infty$, there exists a nontrivial T-periodic solution of

$$\dot{x} = J \nabla h(t,x) .$$

If moreover

$$(1.2) \qquad |j_\circ - j_\infty| > 2n + 1 ,$$

then there are at least two nontrivial T-periodic solutions.

Remark. 1) The conclusion about the existence of two forced oscillations, whenever the purely topological assumption (1.2) is satisfied, stresses the analogy of the result with the P.-B. theorem; the minimal necessary difference between $j_\circ$ and j_∞ is measured by an intrinsic number: 2n +1.

2) From a practical point of view it is a big advantage to be able to deduce the second nontrivial T-periodic solution, without requiring the nondegeneracy of the first one. This assumption is in fact very difficult, if not impossible, to verify.

3) The theorem reported from the quoted paper of Conley and Zehnder appears actually there as a Corollary of a Theorem, which (under additional hypotheses) also gives additional statements.

4) The approach by means of the generalized Morse theory of Conley, used to solve the mentioned problem, allowed also Conley and Zehnder ([6]) to prove (for the case of 2n tori T^{2n}) a conjecture of V.I. Arnold, concerning fixed points of symplectic maps on compact symplectic manifolds. For recent work in this direction we point out [7] and the bibliography there.

2. SKETCH OF THE PROOF

1) the first step, according to a procedure which has often been used in this conference, consists of reformulating the problem of finding T-periodic solutions of (1.1) as a variational one. Namely it con-

sists of looking equivalently for critical points of a related functional on an opportune function space.

We recall the main points in this direction, following the work of Conley and Zehnder ([5]).

Let $H = L^2(0,T;R^{2n})$ and define in H the linear operator

$A: dom(A) \subset H \to H$ by setting

$$dom(A) = \{u \in H^1(0,T;R^{2n}) : u(T) = u(0)\}$$

as

$$Au \doteq = J\dot{u} \qquad u \in dom(A) .$$

Then define the continuous operator

$F: H \to H$ by

$$F(u)(t) \doteq \nabla h(t,u(t)) .$$

Writing (1.1) in the form

$$-J\dot{x} = \nabla h(t,x) ,$$

we immediately see that every solution u of the equation

$$\text{(2.1)} \qquad Au = F(u)$$

defines a classical T-periodic solution of (1.1). And conversely a T-periodic solution of (1.1) defines (by restriction) a solution of (2.1).

Now: equation (2.1) is the Euler equation of the variational problem

$$\text{extr } \{f(u) : u \in dom(A)\}$$

where

$$f(u) = \frac{1}{2} \langle Au, u \rangle_H - \Phi(u) ,$$

$$(u) = \int_0^T h(t,u(t))dt$$

with periodic boundary conditions: u(T) = u(0).

This functional is not bounded from below nor from above. Because of this, the techniques of standard variational calculus are not directly applicable.

2) But it turns out in our case that the assumption

$$(2.2) \qquad -\beta \leq h_{xx}(t,x) \leq \beta \qquad (t,x) \in R \times R^{2n}, \quad \beta > 0$$

allows to reduce the problem to the study of critical points of a related function g, defined on a finite dimensional space Z.

If $\{E_\lambda : \lambda \in R\}$ denotes the spectral resolution of the operator A,, which is self-adjoint and has a pure point spectrum, and if P denotes the orthogonal projection given by

$$P = \int_{-\beta}^{\beta} dE_\lambda$$

(ß must not belong to the spectrum of A), then Z = PH. (Assumption (2.2) implies that the nonlinearity F interacts only with finitely many eigenvalues of A; those contained in the interval $(-\beta, \beta)$. See [1], [2] and [3]).

3) Therefore the problem is now reduced to look for critical points of the gradient flow

$$(2.3) \qquad \dot{x} = \nabla g(x) \qquad x \in Z.$$

Here is that the Morse-type index theory developed by Conley is used.

From the assumption that the linearized Hamiltonian System at infinity, $\dot{y} = JA\nabla(t)y$, is nondegenerate, one deduces (see ([5]) that the set S of bounded orbits of (2.3) is compact. It has hence a Conley index, h(S). This can be computed, using invariance properties of the Conley index under deformations, and it turns out to be the homotopy-type of a pointed sphere:

$$h(S) = [\dot{S}^{m_\infty}] \quad .$$

$\dot{S}^{m_\infty}$ denotes a sphere of dimension m_∞ with a distinguished point *, i.e. a pair $(S^{m_\infty},*)$, where

$$m_\infty = \frac{1}{2} \dim Z - j_\infty .$$

It is hence related to the winding number of the system at infinity.

The critical points x_j of the gradient system (2.3) clearly constitute a Morse decomposition (see [5]) of S. Therefore the following identity holds true:

$$(2.4) \qquad \sum_j p(t,h(\{x_j\})) = p(t,h(S)) + (1+t)\, Q(t),$$

where the algebraic invariant $p(t,h(\{x\}))$ is defined as

$$p(t,h(\{x\})) = \Sigma\, \gamma_j t^j \qquad \nu_j = \text{rank } H_j(N_1,N_0) ;$$

(N_1,N_0) is an "index pair" of x and Q(t) a formal power series having nonnegative coefficients (see [5] for definitions and proof).

It is also proved in [5] that, if a periodic solution is nondegenerate with index j, then the corresponding critical point x of g in Z is an isolated invariant set with index

$$h(\{x\}) = [\dot{S}^m] ,$$

where $m = \frac{1}{2} \dim Z - j$.

Therefore $p(t,h(\{x\})) = t^m$.

According to this, we know that the contribution on the left hand side of (2.4) of the critical point corresponding to the trivial solution $x \equiv 0$, is t^{m_0}, where

$$m_0 = \frac{1}{2} \dim Z - j_0 .$$

The existence of a nontrivial T-periodic solution of (1.1), $x^*(t)$ provided $j_\circ \neq j_\infty$, is now immediately seen, observing that, if the only periodic solution would be $x = 0$, then it would be

$$t^{m_\circ} = t^{m_\infty} + (1+t)\, Q(t) ,$$

where $Q(t)$ is a polynomial with nonnegative integer coefficients. But this is absurd.

4) What allows now to obtain our result are the two following observations:

Lemma 1. If $\hat{x}$ is a critical point of g, then

$$\dim \operatorname{Ker} d^2g(\hat{x}) \leq 2n .$$

Lemma 2. If $\hat{x}$ is a (possibly degenerate) critical point of g and λ denotes its classical Morse index, then

$$p(t,h(\{\hat{x}\})) = \sum_{j=0}^{\lambda+2n} a_j t^j \qquad a_j \geq 0$$

Taking into account Lemma 2, the result can be deduced by contradiction: if x^* would be the only nontrivial critical point of g, the identity (2.4) could be written:

$$(2.5) \qquad t^{m_\circ} + t^{\lambda} \sum_{j=\lambda}^{2n} a_j t^j = t^{m_\infty} + (1+t)\, Q(t) ,$$

with

$\lambda \leq \min \{ m_\circ + 1, m_\infty \}$. (The last estimate can easily be deduced reasoning by contradiction).

Now: (recall we are supposing $j_\circ \neq j_\infty$, and therefore $m_\circ \neq m_\infty$) suppose e.g. that it is $m_\circ > m_\infty$. (The case $m_\infty > m_\circ$ can be worked out analogously).

We have then

$$(2.6) \qquad \lambda \leq m_\infty .$$

On the r.h.s. of (2.5) there must be a $t^{m_\circ}$ and hence also a $t^{m_\circ -1}$ or a $t^{m_\circ +1}$. A $t^{m_\circ -1}$ or a $t^{m_\circ -1}$ (the same on the r.h.s.) must be also on the l.h.s. . But in view of (2.6) this is impossible, when the estimate (1.2) holds true. In the present situation in fact (1.2) is equivalent to:

$$m_\circ - m_\infty > 2n + 1 .$$

Conclusion: the assumption that x* is the only nontrivial critical point must be wrong: there is therefore a second nontrivial critical point.

We refer for more detailed proof and for additional statements to [11] (see also [12]).

For time independent Hamiltonian Systems (where clearly richer multiplicity results are expected) recent results are contained in [13], where also subharmonic solutions are considered.

We should also mention that, for the autonomous case, multiplicity results were already contained in [3], and in [14] and [15]; in these papers minimax techniques for functionals in presence of symmetry were used.

Existence of two forced oscillations for certain second order Hamiltonian Systems was obtained in [8] . For the specific hypotheses needed, we refer the reader to that paper.

REFERENCES

[1] Amann, H.: "Saddle points and multiple solutions of differential equations", Math. Z., 169, (1979), pp. 127-166

[2] Amann, H., Zehnder, E.: "Nontrivial solutions for a class of non-linear differential equations", Annali Sc. Norm. Sup. Pisa, Serie IV, Vol. VII, (1980), pp. 593-603

[3] Amann, H., Zehnder, E.: "Periodic solutions of asymptotically linear Hamiltonian Systems", Manus. Math., 32, (1980), pp. 149-189

[4] Conley, C.C.: "Isolated invariant sets and the Morse index", CBMS Regional Conf. Series in Math., 38, (1978), AMS Providence R.I.

[5] Conley, C.C., Zehnder, E.: "Morse type index theory for flows and Periodic solutions of Hamiltonian Equations", Comm. Pure and Appl. Math., Vol. XXXVII, (1984), pp. 207-253

[6] Conley, C.C., Zehnder, E.: "The Birkhoff-Lewis fixed point theory and a conjecture of V.I. Arnold", Invent. Math., 73, (1983), pp. 33-49

[7] Floer, A.: "Proof of the Arnold conjecture for surfaces and generalizations for certain Kähler manifolds", Duke Math. Journ., Vol. 53, n. 1, (1986), pp. 1-32

[8] Coti Zelati, V.: "Perturbations of second order hamiltonian systems via Morse theory", Boll. U.M.I., Anal. Funz. e Appl., serie VI, Vol. IV-C, n. 1, (1985), pp. 307-322

[9] Gromoll, D., Meyer, W.: "On differentiable functions with isolated critical points", Topology, 8, (1969), pp. 361-369

[10] Marino, A., Prodi, G.: "Metodi perturbativi nella teoria di Morse", Boll. U.M.I., (4), 11, Suppl. fasc. 3, (1975), pp. 1-32

[11] Bertotti, M.L.: "Forced Oscillations of Asyptotically linear Hamiltonian Systems", to appear in Boll. U.M.I.

[12] Bertotti, M.L., Zehnder, E.: "A Poincaré-Birkhoff type result in higher dimensions", Proceedings of the Int. Conf. "Stochastic Processes in Classical and Quantum Systems", June 1985, Ascona, Switzerland

[13] Zehnder, E.: "A Poincaré-Birkhoff type result in higher dimensions", Preprint, Bochum (1986)

[14] Benci, V.: "On critical point theory for indefinite functionals in the presence of symmetries", Trans. A.M.S., Vol. 274, n.2,

(1982), pp. 533-572

[15] Benci, V., Capozzi, A., Fortunato, D.: "Periodic solutions of Hamiltonian Systems with prescribed period", M.R.C. Technical Summary Report = 2508.

GENERATING PHASE FUNCTIONS AND HAMILTONIAN SYSTEMS

Marc Chaperon

Centre de mathématiques - Ecole Polytechnique
91128 Palaiseau Cedex (France)
"U. A. du CNRS n° 169"

The aim of this article is to present recent work of J.C. Sikorav (S 85b) on the Arnold's conjectures in symplectic geometry (Ch 83), providing extremely simple proofs of results which were out of reach four and even two years ago. This simplicity was attained by successive approximations ((CZ 82), (Ch 84), (LS 85), S 85a)). Of course, we give applications to periodic orbits of hamiltonian systems. However, the reader is referred to (Ch 84, Théorème 2) for a more direct approach, based on a previous state of the same ideas.

As this is an introductory paper, we shall not strive for the greatest possible generality; thus, the only compact manifolds considered will be tori - a nice feature of Sikorav's approach is that the general case follows from that of the torus by routine modifications, whereas the similar passage from (Ch 83) to (H 84) was really hard and that from (Ch 84) to (LS 85) (which contains (H 84) and is contained in (S 85b)) demanded an additional idea.

All functions are assumed smooth.

1. GENERATING PHASE FUNCTIONS

Let M denote the n-torus $\mathbb{R}^n/\mathbb{Z}^n$. A <u>phase function on</u> M is a function $F : M \times E \to \mathbb{R}$, where E denotes a finite dimensional real vector space. Given such an F, let $d_H F : M \times E \to M\ (\mathbb{R}^n)^* = T^*M$ and $d_V F : M \times E \to E^*$ be given by $d_H F(q,v) = (q, \partial F/\partial q(q,v))$ and $d_V F = \partial F/\partial v(q,v)$. If $0 \in E^*$ is a regular value of $d_V F$ (in which case F is called <u>non-degenerate</u>), then $\Sigma_F = (d_V F)^{-1}(0)$ is a submanifold of $M \times E$, and the mapping $j_F := d_H F | \Sigma_F : \Sigma_F \to T^*M$ is an immersion; moreover, this immersion is <u>exact lagrangian</u>, i.e. the 1-form $j_F^*(pdq)$ is exact, where pdq denotes the <u>Liouville form</u> of T*M (which is defined by the fact $\alpha^*(pdq) = \alpha$ for every 1-form α on M, viewed as a mapping $M \to T^*M$ on the left-hand side of this identity). An (exact lagrangian) immersion j of some manifold L into T*M is <u>generated by a phase function</u> if $j = j_F \circ g$ for some <u>non-degenerate</u> phase function F on M and some diffeomorphism $g : L \to \Sigma_F$.

Following (S 85b), call a phase function $F : M \times E \to \mathbb{R}$ <u>quadratic</u> if there exists a <u>non-degenerate</u> quadratic form Q on E such that $F(q,v) = Q(v)$ outside some compact subset of $M \times E$. The following - essentially classical - result is the Morse-Lyusternik-Schnirelmann part of the theory (for a proof, <u>see</u> for example (Ch Z 83)):

<u>Proposition 1</u>. <u>A quadratic phase function on</u> M <u>has at least</u> n+1 <u>critical points, and at least</u> 2^n <u>if none of them is degenerate (for a general compact manifold</u> M, <u>these lower bounds can be replaced by</u> 1 + cup-length(M) <u>and by the sum of the Betti numbers of</u> M <u>respectively</u>).

Here is an obvious consequence:

<u>Corollary 1</u>. <u>Let</u> 0_M <u>denote the zero section</u> $M \times \{0\}$ <u>of</u> T*M. <u>If a lagrangian immersion</u> $j : L \to T^*M$ <u>is generated by a quadratic phase function, then</u> $j^{-}(0_M)$ <u>contains at least</u> n+1 <u>points, and at least</u> 2^n <u>if</u> j <u>is transverse to</u> 0_M.

2. A THEOREM OF SIKORAV

Let $I := [0,1]$. Recall that an isotopy (g_t) of T^*M is a path $t \to g_t$, $t \in I$, in the group of diffeomorphisms of T^*M, such that $g_0 = \mathrm{Id}$ and that $(t,x) \to g_t(x)$ is smooth. The family $(\dot{g}_t)$ of vector fields on T^*M given by $\frac{d}{dt} g_t = \dot{g}_t \circ g_t$, $t \in I$, determines the isotopy: it is called the infinitesimal generator of (g_t); an isotopy (g_t) of T^*M is hamiltonian if the following two equivalent conditions hold:

(i) The 1-form $pdq - g_{t*}(pdq)$ is exact for every $t \in I$.

(ii) For each $t \in I$, the vector field $\dot{g}_t$ is hamiltonian, i.e. $\dot{g}_t \lrcorner\, d(pdq) = dH_t$ for some function $H_t : T^*M \to \mathbb{R}$, where $\lrcorner$ is the interior product.

Clearly, the H_t's can be so chosen that $(t,x) \to H_t(x)$ is smooth, in which case the family (H_t) is called a hamiltonian of (g_t).

Theorem 1 (S 85b). If a lagrangian immersion j of some compact manifold L into T^*M is generated by a phase function (resp. by a quadratic phase function), so is $g_1 \circ j$ for every hamiltonian isotopy (g_t) of T^*M.

Proof. As L is compact, multiplying a hamiltonian of (g_t) by a suitable bump function on T^*M, we may and shall assume that (g_t) has compact support, i.e. that every g_t equals the identity outside a fixed compact subset of T^*M.

For each positive integer N, we have that

$$g_1 = h_N \circ \ldots \circ h_0 \, , \text{ where } h_k := g_{(k+1)/(N+1)} \circ g^{-1}_{k/(N+1)} \, .$$

Clearly, for each k, $h = h_k$ satisfies

(1) $pdq - h_*(pdq) = df$, which has compact support, for some $f : T^*M \to \mathbb{R}$

Moreover, if N is large enough, each $h = h_k$ is small in the following sense: setting

$$(2) \qquad (q'(q,p,x),p'(q,p,x)) = h^{-1}(q+x,p) \text{ for } (q,p) \in T^*M \text{ and } x \in \mathbb{R}^n,$$

we have that

$$(3) \qquad \text{every } \frac{\partial q'}{\partial x}(q,p,x) \text{ is an isomorphism of } \mathbb{R}^n \text{ onto itself.}$$

Thus, Theorem 1 can be obtained by applying the following result first to j with $h:=h_o$, then to $j:=h_o \circ j$ with $h:=h$, and so on:

Theorem 2. Let h be a diffeomorphism of T^*M which equals the identity outside a compact set and satisfies (1) and (3). If a lagrangian immersion $j : L \to T^*M$ is generated by a phase functions $F : M \times E \to \mathbb{R}$, then the following hold:

(i) $h \circ j$ is generated by the phase function $G: M \times (\mathbb{R}^n)^* \times \mathbb{R}^n \times E \to \mathbb{R}$ given by

$$G(q,p,x,v) = -px + f(q+x,p) + F(q'(q,x,p),v) \quad ,$$

where f and q' are as in (1)-(2).

(ii) If F is quadratic, $F(q,v) = Q(v)$ outside a compact set, and if R is the non-degenerate quadratic form on $(\mathbb{R}^n)^* \times \mathbb{R}^n \times E$ defined by $R(p,x,v)=-px + Q(v)$, then the mapping

$$M \times ((\mathbb{R}^n)^* \times \mathbb{R}^n \times E) \quad (q,V) \to d_V G(q,V) - dR(V) \quad ((\mathbb{R}^n)^* \times \mathbb{R}^n \times E)^*$$

is bounded. Therefore, if $u : (\mathbb{R}^n)^* \times \mathbb{R}^n \times E \to I$ denotes a compactly supported function, equal to 1 in a neighbourhood of 0, then, for each small enough positive constant c, the phase function $(q,V) \to R(V) + u(cV)(G(q,V) - R(V))$ is quadratic and generates $h \circ j$.

Proof. By (1) - (2), we have that $d(f(q+x,p)) = pd(q+x)-p'\,dq'$, hence

$$(4)\qquad dG = -x\,dp + p\,dp + (\frac{\partial F}{\partial q'}(q',v)-p')dq' + d_V F(q',v)dv .$$

Therefore, by (3), we have that $(q,p,x,v \in \Sigma_G$ if and only if $x=0$ and $d_V F(q',v) = 0$ and $p' = \frac{\partial F}{\partial q'}(q',v)$, i.e. if and only if $(q',v)\ \Sigma_F$ and $(q',p') = j_F(q',v) = h^{-1}(q,p)$ and $x = 0$. It follows that $(q,p,0,v) \to (q'(p,0,v),v)$ is a diffeomorphism of Σ_G onto Σ_F and that G generates $h \circ j_F$. So far, we have not proven that G is non-degenerate, but this is obvious: by (3)-(4), G is non-degenerate if and only if the mapping $(q,p,x,v) \to (-x, \frac{\partial F}{\partial q'}(q',v)-p', d_V F(q',v))$ admits 0 as a regular value, which is the case because $(q,p,x,v) \to (q',p',x,v)$ is a diffeomorphism and F is a nondegenerate phase function.

Under the hypotheses and with the notations of (ii), for each $(q,V) = (q,p,x,v) \in M \times (\mathbb{R}^n)^* \times \mathbb{R}^n \times E$, we have that

$$d_V G(q,V) - dR(V) = d(f(q+x)) + d(F(q',v) - Q(v)) .$$

This is bounded, since both df and $(q,v) \to F(q,v) - Q(v)$ have compact support and dq' is bounded. As the rest of (ii) is straight-forward, this proves Theorem 2 and hence Theorem 1.

3. CONSEQUENCES

Corollary 2. For every hamiltonian isotopy (g_t) of T*M, there are at least n+1 points in $0_M \cap g_1(0_M)$, and at least 2^n if all these intersections are transversal.

Proof. The canonical embedding of M as 0_M is generated by a quadratic phase function, namely ... the zero function on $M \times \{0\}$ - the reader can check that, in our context, it is quite reasonable to consider that the sole quadratic form on $E := 0$ is non-degenerate. Therefore, we can apply Theorem 1 and Corollary 1.

In the sequel we assume $n = 2k$ for some positive integer k. Recall that the standard symplectic form σ on $\mathbb{C}^k$ is the 2-form defined by $\sigma(z)(v,w) = \mathcal{I}m(v.\bar{w})$ for $z,v,w \in \mathbb{C}^k$, where $v.\bar{w}$ is the standard hermitian product and $\mathcal{I}m$ denotes the imaginary part. Viewing M as $\mathbb{C}^k/(\mathbb{Z} + i\mathbb{Z})^k$, the standard symplectic form ω on M is the 2-form obtained from σ via the canonical projection $\mathbb{C}^k \to M$.

Proposition 2 (Ch Z 83). Let Ω be the symplectic form on $M \times M$ given by $\Omega = pr_2^*\omega - pr_1^*\omega$, where $pr_1, pr_2 : M \times M \to M$ denote the projections. There exists a covering projection $P : T^*M \to M \times M$ such that $P^*\Omega = d(pdq)$ and $P|0_M$ is a diffeomorphism onto the diagonal ΔM.

Of course, an isotopy (g_t) of M is called hamiltonian if the 1-form $\dot{g}_t \lrcorner\, \omega$ is exact for every t.

Corollary 3 (CZ 82). For every hamiltonian isotopy (f_t) of M, the number of fixed points of f_1 is at least $2k+1$, and at least 2^{2k} if none of them is degenerate (see (Ch Z 83), p. 98, for a rephrasing in terms of periodic orbits).

Proof. Clearly, one defines a hamiltonian isotopy (F_t) of $M \times M$ (for the symplectic form Ω) by $F_t(x,y) = (x, f_t(y))$. Since P has the unique path lifting property, there exists a unique isotopy (g_t) of T^*M such that $P \circ g_t = F_t \circ P$ for every t. As the properties of P stated in Proposition 2 imply that (g_t) is hamiltonian and that $P|(0_M \cap g_1(0_M))$ is an injection into the intersection of ΔM with the graph of f_1, the number of fixed points of f_1 is at least $\#(0_M \cap g_1(0_M))$, which has the required lower bound by Corollary 2.

Note. Theore 1 is true for an arbitrary compact manifold M, as already mentioned. This is not the case of Proposition 2 in general, but (S 85b) proves it for an interesting class of symplectic manifolds, including all surfaces of positive genus - here again, the proof if very simple -, hence Corollary 3 for these manifolds. Apparently, the idea of using Theorem 1 to prove Corollary 3 is a good idea, which should work in more general situations - but this is work in progress.

REFERENCES

(CZ 82) Conley, C.C., Zehnder, E.: "The Birkhoff-Lewis fixed point theorem and a conjecture of V.I. Arnol'd", Inv. Math. 73 (1983), 33-49

(Ch 83) Chaperon, M.: "Quelques questions de géométrie symplectique", Séminaire Bourbaki 1982-83, Astérisque 105-106 (1983), 231-249

(Ch Z 83) Chaperon, M., Zehnder, E.: "Quelques résultats globaux en géometrie symplectique", Géométrie symplectique et de contact: autour du théorème de Poincaré-Birkhoff (P. Dazord, N. Desolneux-Moulis ed.), Travaux en cours, Hermann, Paris (1984), 51-121

(Ch 84) Chaperon, M.: "Questions de géométrie symplectique", Géométrie symplectique et mécanique (J.P. Dufour ed.), Travaux en cours, Hermann, Paris (1985), 30-45

(H 84) Hofer, H.: "Lagrangian embeddings and critical point theory" preprint, University of Bath (1984)

(LS 85) Laudenbach, F., Sikorav, J.C.: "Persistance d'intersection avec la section nulle au cours d'une isotopie hamiltonienne. ..,"Inv. Math. 82 (1985), 349-357

(S 85a) Sikorav, J.C., "Sur les immersions lagrangiennes dans un fibré cotangent admettant une phase génératrice globale", C. R. Acad. Sci. Paris (1986)

(S 85b) Sikorav, J.C.: "Problèmes d'intersections et de points fixes en géométrie hamiltonienne", preprint, Orsay (1985).

PERIODIC SOLUTIONS OF SECOND ORDER HAMILTONIAN SYSTEMS AND MORSE THEORY

Vittorio Coti Zelati (*)

International School for Advanced Studies
Strada Costiera 11, 34014 Trieste
ITALY

0. INTRODUCTION

The purpose of this paper is to prove existence of T-periodic solutions for the following systems of ordinary differential equations:

$$(V) \qquad -\ddot{x} = \nabla_x V(t,x) ,$$

where $V \in C^2(\mathbb{R} \times \mathbb{R}^N; \mathbb{R})$, $V(t+T,x) = V(t,x)$ for every $(t,x) \in \mathbb{R} \times \mathbb{R}^N$.

This problem has been studied by many authors; we refer here to [Be] and [Ra] for surveys of the topic.

In this paper we will show how Morse theory can be used to prove existence of two T-periodic solutions of (V) under suitable assumptions on the behaviour of V for x close to zero and for x large.

The paper is organized as follows: in § 1 we prove an abstract theorem on the existence of critical points for a C^2 functional defined in a Hilbert space H. This theorem follows directly from classical Morse inequalities using a result by Marino and Prodi [MP] on degenerate critical points. In § 2, we apply the abstract theorem to

(*) Supported by Ministero P.I. Gruppo Nazionale "Calcolo delle Variazioni" (40%)

two different situations, proving in both cases the existence of two nonzero T-periodic solutions for (V). These results are related to previous work of the A. (see [CZ1], [CZ2]).

1. THE ABSTRACT RESULT

Let H be a Hilbert space, and $f : H \to \mathbb{R}$. We define the sets f^c, f_c as $f^c = \{x \in H: f(x) \le c\}$, $f_c = \{x \in H: f(x) \geq c\}$. Moreover, we will say that $f \in C^1(H;\mathbb{R})$ satisfies the Palais-Smale (PS) condition in the set A if for every sequence $x_n \in A$ such that $f(x_n)$ is bounded and $f'(x_n) \to 0$ one has a subsequence x_{n_k} converging to $\bar{x}$ A with $f'(\bar{x}) = 0$.

We will denote with $H_*(A,B)$, where $B \subset A \subset H$, the homology of the couple (A,B). We are now in position to state:

Theorem 1. Suppose $f \in C^2(H;\mathbb{R})$ satisfies (PS) in f_c and that $f'(x)$ is Fredholm of index 0 for every $x \in H$.

1°- If

i) $H_*(H,f^c) \neq \{0\}$,

then exists $\bar{x} \in H$ such that $f'(\bar{x}) = 0$.

2°- If i) holds and, moreover

ii) $\dim \ker d^2f(x) \leqq m \qquad \forall\, x \in H$;

iii) $H_q(H,f^c) = \{0\} \qquad \forall\, q \geqq \bar{q}$;

iv) $f'(0) = 0$ and 0 is a non-degenerate critical point of Morse index $\lambda \geqq \bar{q} + m + 1$,

then f has at least two nonzero critical points.

Proof. To prove 1°- we suppose, by contradiction, that f has no critical points in f_c. Then, since in f_c (PS) holds, it is well known (see for example [MP]) that f^c is a deformation retract of H. Hence

$H_*(H,f^c) = 0$, contradiction with i) which proves 1°-.

To prove 2°- we begin assuming, by contradiction, that 0 is the only critical point for f in f_c.

Setting

$$(1) \qquad R_k = \dim H_k(H,f^c)$$

we have that the Poincaré polinomial of the couple (H,f^c) is given by (using iii)):

$$(2) \qquad \mathcal{P}(t) = \sum_{k=0}^{\bar{q}-1} R_k\, t^k$$

while our assumption implies that the Morse polinomial of f is given by:

$$(3) \qquad \mathcal{M}_f(t) = t^\lambda .$$

The Morse inequalities can be put in the form (see [Bo])

$$(4) \qquad \mathcal{M}_f(t) = \mathcal{P}(t) + (1+t)Q(t)$$

where $Q(t) = a_o+a_1t+\ldots+a_nt^n+\ldots$ with $a_i \geqq 0$. Using (2) and (3) one deduces from (4)

$$t^\lambda = \sum_{k=0}^{\bar{q}-1} R_k\, t^k + (1+t)Q(t) .$$

Since, by i) and iii), exists $q' < \lambda$ such that $R_{q'} \neq 0$ and $a_i \geqq 0$ for every i, we reach a contradiction which proves that 0 cannot be the only critical point of f in f_c.

We now suppose that f has only two critical points, 0 and $\bar{x}$. We know that 0 is a nondegenerate critical point of index λ, while $\overline{x}$ is (eventually) degenerate. Let $\overline{\lambda}$ be the Morse index of $\overline{x}$. Using an argument of [MP] (see also [GM]), one can prove as in [CZ1] that its contribution to the Morse polinomial of f (thanks to the assumption ii)) is of the form (see also [Br])

$$t^{\overline{\lambda}}(b_o+b_1t+\ldots+b_mt^m)$$

so that, in this case, the Morse polinomial of f is

$$\mathcal{M}_f(t) = t^\lambda + t^{\overline{\lambda}}(b_o+b_1t+\ldots+b_mt^m) . \tag{5}$$

Using (5) and (2) in (4), we deduce

$$t^\lambda + t^{\overline{\lambda}}(b_o+b_1t+\ldots+b_mt^m) = \sum_{k=0}^{\overline{q}-1} R_k t^k + (1+t)Q(t) . \tag{6}$$

Let $\overline{k} = \inf \{k \in \mathbb{N} : R_k \neq 0\} < \overline{q}$. (6) implies that $\overline{\lambda} \leqq \overline{k} < \overline{q}$, so that $\overline{\lambda}+m < \overline{q}+m$. Considering (6) up to the power $\overline{q}+m-1 < \lambda$ we deduce

$$t^{\overline{\lambda}}(b + b\ t+\ldots b_mt^m) = \sum_{k=0}^{\overline{q}-1} R_k t^k + \sum_{i=0}^{\overline{q}+m-1} (a_i+a_{i+1})t^i ,$$

so that (6) becomes:

$$t^\lambda = \sum_{i=\overline{q}+m}^{\infty} (a_i+a_{i+1})t^i \tag{7}$$

since $\lambda > \overline{q} + m$, we have, for every $i \neq \lambda$, $a_i+a_{i+1} = 0$, which implies $a_i = 0$ for every $i \geqq \overline{q}+m$, contradiction with (7) which proves our theorem.

2. SOME APPLICATIONS

Theorem 1, both part 1°- and 2°-, has many applications when one is looking for critical points of a functional. We will here deal with two applications to the problem of finding T-periodic solutions of

$$(8) \qquad -\ddot{x} = \nabla_x V(t,x)$$

where $V \quad C^2(\mathbb{R} \times \mathbb{R}^N; \mathbb{R})$, $V(t+T,x) = V(t,x)$.

The following theorem is related to theorem 1.7 of [CZ2].

Theorem 2. Suppose $V \in C^2(\mathbb{R} \times \mathbb{R}^N; \mathbb{R})$, $V(t+T,x) = V(t,x)$ for every $(t,x) \quad \mathbb{R} \times \mathbb{R}^N$. We also assume:

(V1) $0 \leqq V(t,x) < m \qquad \forall (t,x) \in \mathbb{R} \times \mathbb{R}^N$, $V(t,x) \to m$ uniformly in t as $|x| \to +\infty$, monotonically increasing along rays as $|x|$ large

(V2) $\nabla_x V(t,x) \to 0$ uniformly in t as $|x| \to +\infty$.

Then (8) has at least one T-periodic solution.

If, moreover

(V3) $\nabla_x V(t,0) = 0 \quad \forall t \in \mathbb{R}$, and $V''_{xx}(t,0) = \beta I_N$ with $\beta > \frac{16\pi^2}{T^2}$, $\beta \neq \frac{k^2 4\pi^2}{T^2}$ for every $k \in \mathbb{N}$,

then (8) has at least two nonzero solutions.

Proof. We will apply theorem 1. Let $H = H^1(S^1;\mathbb{R}^N)$, $f(x) = \frac{1}{2}\int_0^T |\dot{x}|^2 - \int_0^T V(t,x)$. f is clearly of class C^2; moreover $f'(x)$ is of the form identity minus compact, hence Fredholm of index zero. It satisfies the (PS) condition (see [CZ2], lemma 1.2) in $f_{-Tm+\varepsilon}$ for every $\varepsilon > 0$. As in [AC], one can evaluate $H_*(H,f^{-Tm+\varepsilon})$ for $\varepsilon > 0$ small enough (it is not restrictive to assume that there are no critical points in $f^{-Tm+\varepsilon}$). One finds:

$$(9) \qquad H_*(f^{-Tm+\varepsilon}) = H_*(S^{N-1}) \ .$$

Then, from the reduced homology sequence of the couple $(H,f^{-Tm+\varepsilon})$ one deduces

$$\to \tilde{H}_{q+1}(H) \to H_{q+1}(H,f^{-Tm+\varepsilon}) \to \tilde{H}_q(f^{-Tm+\varepsilon}) \to \tilde{H}_q(H) \to ;$$

since $H_q(H) = \{0\}$, using (9) one deduces

$$(10) \qquad H_{q+1}(H,f^{-Tm+\varepsilon}) = \tilde{H}_q(S^{N-1}).$$

In particular

$$(11) \qquad H_N(H,f^{-Tm+\varepsilon}) \neq \{0\},\ H_q(H,f^{-Tm+\varepsilon}) = \{0\} \text{ for every } q \neq N$$

and i) of theorem 1 holds, as well as iii) with $\bar{q} = N + 1$.

We can now apply 1°- of theorem 1 and find a solution of (8). The second statement of the theorem will follow from 2°- of theorem 1. We still has to verify that ii) and iv) hold.

ii) follows upon noticing that $v \in \ker d^2f(x)$ iff $-\ddot{v}-V''_{xx}(t,x(t))v=0$ this implies that $\ker d^2f(x)$ is, at the most, 2N dimensional.

To show that also iv) holds, set, for $v \in H$,

$$v(t) = \sum_k v_k \exp(i \frac{2\pi}{T} kt) \ .$$

Then

$$d^2f(0)[v,v] = T \sum_k (\frac{k^2 4\pi^2}{T^2} - \beta)|v_k|^2 \ .$$

From (V3) immediately follows 0 is a nondegenerate critical point and that its index is greater than 5N, hence also iv) holds.

In an analogous way it can be proved the following

Theorem 3. Let $V: \mathbb{R} \times \mathbb{R}^N \to \mathbb{R}$ be such that:

(V1) $V \in C^2(\mathbb{R} \times \mathbb{R}^N; \mathbb{R})$, $V(t,0) = 0$, $V(t+T,x) = V(t,x)$

(V2) $V(t,x) = \frac{1}{2} k\ |x|^2 + U(t,x)$ where $k \in]0,1[$ and $|U_x(t,x)| \leqq \phi(|x|)$ where $\phi(s)/s \to 0$ as $s \to +\infty$.

(V3) $\nabla_x V(t,0) = 0$, $V''_{xx}(t,0) = \beta I_N$ with $\beta > \frac{16\pi^2}{T^2}$, $\beta \neq \frac{n^2 4\pi^2}{T^2}$ for every $n \in \mathbb{N}$.

Then (8) has at least two nonzero solutions.

Proof. See theorem 3.7 (ii) of [CZ1].

REFERENCES

[AC] Ambrosetti, A., Coti Zelati, V.: "Periodic solutions of dynamical systems with singular potential", preprint 1986

[Be] Berestycki, H.:"Solutions periodiques de systemes hamiltoniens" Université Pierre et Marie Curie (Paris VI), 1983

[Bo] Bott, R.: "Lectures on Morse theory, old and new", Bull. Am. Math. Soc. 7, (1982), pp. 331-358

[Br] Bertotti, M.L.: "Forced oscillations of asymptotically linear Hamiltonian systems", preprint SISSA 20/86/MP, Trieste 1986

[CZ1] Coti Zelati, V.: "Perturbations of second order Hamiltonian systems via Morse theory", Bollettino UMI, serie VI, vol IV-C, (1985), pp. 307-322

[CZ2] Coti Zelati, V.: "Periodic solutions of dynamical systems with bounded potential", to appear on Jour. Diff. Equations

[GM] Gromoll, D., Meyer, W.: "On differentiable functions with isolated critical points", Topology, 8 (1969), pp. 361-369

[MP] Marino, A., Prodi, G.: "Metodi perturbativi nella teoria di Morse", Boll. UMI (4), 11, Suppl. fasc. 3 (1975), pp. 1-32

[Ra] Rabinowitz, P.H.: "Periodic solutions of hamiltonian systems: a survey", SIAM J. Math. Anal., 13 (1982), pp. 343-352.

PERIODIC SOLUTIONS WITH BOUNCING OF HAMILTONIAN PROBLEMS AND THEIR MINIMAL PERIODS

F. Giannoni

Dipartimento di Matematica II Università di Roma

About the study of periodic trajectories of a material point in a potential field there are many works. An important chapter of this work concerns the case in which it is given the minimal period of the trajectory (see 1,3,5,6).

The problem that I expose in this seminar concerns the periodic trajectories with prescribed minimal period when the material point moves in a potential field and bounces with an elastic collision against the boundary of a convex billiard in R^n.

The aim is the study of the properties of such which trajectories and, particularly, the number of those which differ geometrically and have prescribed minimal period.

I don't suppose the billiard to be regular. More exactly it is given an open convex set Ω in R^n without hypothesis of regularity, the potential energy V and the period T.

I consider which periodic orbits (which I call principal bounce trajectories) $q:R \to \bar{\Omega}$ having period T, nonconstant and such that: $q(0) \in \partial\Omega$, $q(T/2) \in \partial\Omega$, q is of class C^2 in $[0,T/2]$ and $[T/2,T]$ and verifies $\ddot{q} + \text{grad } V(q) = 0$, $-\lim_{t\to 0^+} q(t) = \lim_{t\to 0^-} q(t) \equiv q(0^-)$, $-\lim_{t\to T^+/2} q(t) = \lim_{t\to T^-/2} q(t) \equiv q(T/2^-)$, $-q(0^-)$ and $q(T/2^-)$ are in the cone of the normal directions to the convex $\bar{\Omega}$ respectively in $q(0)$ and $q(T/2)$.

I suppose everywere that the force -gradV be not "towards" Ω in the points of $\partial\Omega$ and that the convexity of V be controlled.

Proofs of results that are listed in this seminar, are given in a forthcoming paper.

Let Ω be an open bounded convex set in R^n, and let $V \in C^2(R^n,R)$.

Definition 1. A curve $q:R \to \bar{\Omega}$ is called principal bounce trajectory in $\bar{\Omega}$ with respect to potential V,T-periodic, if

1) q is continuous, T-periodic, and nonconstant.

2) In [0,T/2] and [T/2,T] q is of class C^2 and verifies the equation $\ddot{q} + \mathrm{grad}V(q) = 0$.

3) $q(kT/2) \in \partial\Omega$ for all $k \in Z$.

4) $\dot{q}(0^+) = -\dot{q}(0^-)$, $\dot{q}(T/2^+) = -\dot{q}(T/2^-)$ where $\dot{q}(a^\pm) = \lim_{t\to a^\pm} q(t)$.

5) $-\dot{q}(0^+) \in N_{q(0)}(\bar{\Omega})$, $\dot{q}(T/2^+) \in N_{q(T/2)}(\bar{\Omega})$ where if $x \in \partial\Omega$, $N_x(\bar{\Omega}) = \{e \in R^n: (y-x,e) \le 0$ fo all $y \in \bar{\Omega}$, where (,) is the usual inner product in R^n.

Definition 2. q is said open if $q(0) \neq q(T/2)$ and not open if $q(0) = q(T/2)$.

Remark 3. As gradV is of class C^1, from the properties 1) and 4) we have that $q(-t) = q(t)$ for all t in R.

From 1) - 5) we don't have that T is the minimal period.

- Ω open bounded convex set (also non regular) in R^n,
- $F = -\mathrm{grad}V$ a conservative force field,
- T the period,

are assigned.

We assume:

Hypothesis on Ω : Ω open bounded convex set in R^n.

Hypothesis on F: F is not ingoing $x \in \partial\Omega$, that is for all x in

(*) $\partial\Omega$ we have $(F(x),e) \geq 0$ for all e in $N_x(\bar{\Omega})$.

Hypothesis on T: $1-(c_\Omega T^2)/\pi^2 > 0$ where $c_\Omega = \sup_{x \in \bar{\Omega}} \{$eigenvalues of $V''(x)\}$

We obtain the following results:

Theorem 4. Let us suppose (*) and $F(x) \neq 0$ for all $x \in \partial\Omega$.

Then there are 2 geometrically distinct principal bouncing not open trajectories, having minimal period T/2. Besides $q(T/4+t) = q(T/4-t)$ for all t in R.

This estimate is the best. In fact there are example in which we have only 2 principal bouncing not open trajectories.

Let $2R_o$ be the diameter of Ω, $M = \sup_{x \in \bar{\Omega}} V$ and $m = \inf_{x \in \partial\Omega} V$.

Theorem 5. Let us suppose (*) and $2R_o^2 > T^2(M-m)$. Then there is a principal bouncing open trajectory having minimal period T.

Perhaps the hypothes s $2R_o^2 > T^2(M-m)$ may be removed.

To formulate the following theorem we need a definition.

Definition 6. Let S be a sphere in R^n. We see that S is an inscribed maximal sphere in $\bar{\Omega}$ if $S \subset \bar{\Omega}$ and its diameter is maximal among the diameters of the sphere that are in $\bar{\Omega}$.

We indicate with 2r the diameter of an inscribed maximal sphere.

Theorem 6. Let us suppose (*) and $2r^2 > T^2(M-m)$. Then there are n geometrically distinct principal bouncing open trajectories, having minimal period T.

This estimate is the best. For example if $V = 0$ the trajectories that we seek are the principal cords of the convex $\bar{\Omega}$.

Observation 7. If V is a concave function the preceding condition may be weakened. If V and $\bar{\Omega}$ have a same symmetry center it may be dropped.

Observation 8. If in theorems 4,5 and 6 we add one of the following conditions:

- $(F(x),e) > 0$ for all e in $N_x(\bar{\Omega})$ for all x in $\partial\Omega$, or
- $(e_1,e_2) > 0$ for all e_1,e_2 in $N_x(\bar{\Omega})$ for all x in $\partial\Omega$,

then $q(t) \in \Omega$ for all t in $(0,T/2)$.

The problem is reduced to the research of points that are critical from below (see 7) for a function $h: R^n \times R^n \to R \cup \{+\infty\}$, that is constructed in the following way.

We pose $X(A,B) = \{q \in H^{1/2}(0,T/2,R^n): q(0) = A, q, q(T/2) = B\}$, $f : H^{1/2}(0,T/2,R^n) \to R$, $f(q) = 1/2 \int_0^{T/2} (\dot{q},\dot{q})^2 dt - \int_0^{T/2} V(q)dt$ and $g(A,B) = \min f(q)$ when q belongs to $X(A,B)$.

As the third hypothesis of (*) holds, if we modify in a suitable way the function V in the complementary set of $\bar{\Omega}$, we have that the minimal point is unique and the function g is of class C^2.

Let I be the indicatrix function of the convex $\bar{\Omega} \times \bar{\Omega}$, that is $I(A,B) = +\infty$ if $(A,B) \notin \bar{\Omega} \times \bar{\Omega}$ and $I(A,B) = 0$ if $(A,B) \in \bar{\Omega} \times \bar{\Omega}$.

Finally let h be the function such that $h = -g+I$.

The function h is subdifferential in the points of $\bar{\Omega} \times \bar{\Omega}$ in the sense of the following definition (see 7):

Definition 9. Let $h: H \to R \cup \{+\infty\}$, where H is an Hilbert space. Let u a point in Dom $h = \{u: h(u) \text{ is finite}\}$.

We say subdifferential of h in u the set $\partial^- h(u)$ (eventually void) of the elements α of H such that

$$\liminf_{v \to u} (h(v) - h(u) - (\alpha, v-u))/|v-u| \geq 0$$

As $\partial^- h(u)$ is a convex and closed set we indicate with $\text{grad}^- h(u)$ the element that has the minimum norm.

We study the evolution equation $U' = -\text{grad}^- h(U)$ associated to the function h. As $\partial\Omega \times \partial\Omega$ in invariant with respect to flow of such differential equation we can "count" the points that are stationary from below for the function h (that is such that $\text{grad}^- h(u) = 0$) and belong to $\partial\Omega \times \partial\Omega$. These points identify solutions for our problem.

We use a technique of the type of Lusternick and Schnirelman applying the theory of Brezis on the maximal monotone operators in Hilbert spaces (see 2) and the theory of De Giorgi - Marino - Tosques on the curves of maximal slope in metric spaces (see 4,7).

The problems I had to face are the following:

I) To prove that if q in X(A,B) is such that $f(q) = g(A,B)$, then q does not "go out" of $\overline{\Omega}$, if A and B are in $\overline{\Omega}$.

II) To consider the non regularity of $\overline{\Omega}$.

III) To find conditions so that h has points of stationary from below in $\partial\Omega \times \partial\Omega - \{(A,A) : A \text{ belong to } \partial\Omega\}$.

REFERENCES

[1] Benci, V., Fortunato, D.: "Subharmonic solutions of minimal period of second order differential equations", to appear

[2] Brezis, H.: "Operateurs maximaux monotones dans les espaces de Hilbert", North-Holland Math. Studies (50), Amsterdam-London 1973

[3] Clarke, F., Ekeland, I.: "Hamiltonian trajectories having prescribed minimal period" Comm. Pure and Appl. Math. 33 (1980) 103-116

[4] De Giorgi, E., Marino, A., Tosques, M.: "Problemi di evoluzione in spazi metrici e curve di massima pendenza" Atti Acc. Naz. dei Lincei Rend. CL. Sc. (8), 68 (1980) 180/187

[5] Ekeland, I., Hofer, H.: "Periodic solution with prescribed period for convex autonomous system", preprint Ceremade n. 8421, Paris (1984)

[6] Girardi, M., Matzeu, M.: "Periodic solutions of convex autonomous hamiltonian systems with a quadratic growth at the origin and superquadratic at infinity", preprint Ist. Mat. "G. Castelnuovo Roma, 1985

[7] Marino, A.: "Equazioni di evoluzione e molteplicità di punti critici rispetto ad un ostacolo", to appear on Pitman Research Notes on Math.

DUAL VARIATIONAL METHODS AND MORSE INDEX THEORY FOR PERIODIC SOLUTIONS OF HAMILTONIAN SYSTEMS

Mario Girardi and Michele Matzeu

Dipartimento di Matematica - Istituto G. Castelnuovo
Università di Roma "La Sapienza"
Piazzale Aldo Moro 2 - 00185 ROMA - Italy

1. The object of this paper is to expose some results obtained by the authors about the existence of periodic solutions, having prescribed minimal period, to the Hamiltonian system

$$(H) \qquad J\dot{z} = H'(z) + 1/2 \langle Qz,z\rangle$$

where $J(x,y) = (y,-x)$ $\forall (x,y) \in R^N \times R^N$, Q is the $2N \times 2N$ matrix $Q = \begin{pmatrix} Q_\circ & 0 \\ 0 & Q_\circ \end{pmatrix}$ with $Q_\circ = \begin{pmatrix} \omega_1 & & 0 \\ & \ddots & \\ 0 & & \omega_N \end{pmatrix}$, $0 < \omega_1 \leqq \omega_2 \leqq \dots \leqq \omega_N$ and $H \in C^2(R^{2N};R)$ is strictly convex and has a superquadratic behaviour.

In concerning with this problem, we use some variational techniques based on the use of a suitable version of the duality principle by Clarke and Ekeland ([3]) and of the Morse index theory introduced by Ekeland and Hofer in [4] in order to deal with the "purely superquadratic" case (that is $\omega_1 = \dots = \omega_N = 0$).

2. Let us assume that H satisfies the following conditions:

$$(1) \qquad H(z)\,|z|^{-2} \to 0 \quad \text{as } |z| \to 0$$

$$(2) \qquad \exists r > 0,\ \beta > 2: \langle H'(z),z\rangle \geq \beta H(z) \quad \text{if } |z| > r$$

In general, if we don't assume further conditions on H and Q, it is well known that the periodic solutions of (H) can have bounded minimal periods. For example, it is easy to show that the Hamiltonian system related to the choice of H given by $H(z) = |z|^4$ and $\omega_1 = \ldots = \omega_N = 1$ admits periodic solutions whose minimal periods cannot be greater than 2π (see [10] for more general examples).

Actually, it is possible to show that $2\pi/\omega_N$ is the real bound for minimal periods when conditions (1) and (2) are satisfied. In fact one has the following

Theorem 1 (see [5]). Let all the previous assumptions be satisfied and let H verify (1) and (2). Then, for any $T < 2\pi/\omega_N$, there exists a solution z_T of (H) having minimal period T. Moreover, if H satisfies the further assumptions:

(3) $$H(z) \geq a_1|z|^\beta \quad \forall\, z \in R^{2N}, \quad a_1 > 0$$

(4) $$H(z) \leqq a_2|z|^\beta \quad \forall\, z \in R^{2N} \quad a_2 > 0$$

(5) $$\langle H'(z), z\rangle \geq \beta H(z) \quad \forall\, z \in R^{2N} \quad ,$$

then one has

(6) $$\lim_{T \to 0^+} \left(\inf_{t \in [0,T]} |z_T(t)| \right) = +\infty$$

(7) $$\lim_{T \to (2\pi/\omega_N)^-} \left(\sup_{t \in [0,T]} |z_T(t)| \right) = 0$$

In order to obtain the existence of solutions having minimal period between $2\pi/\omega_N$ and $2\pi/\omega_1$, a simple non-resonance condition among the ω_i's gives an answer to the problem. In fact, one has the following

Theorem 2 (see [6]). Let all the previous assumptions be satisfied and let H verify (3), (4), (5). Then, for any $j \in \{1,\dots,N\}$ such that

$$(8) \qquad \omega_j/\omega_i \in \mathbb{Q} \quad \forall\, i \in \{1,\dots,N\}, \quad i \neq j ,$$

there exists some $\varepsilon_j > 0$ such that, for any $T \in (2\pi/\omega_j - \varepsilon_j,\ 2\pi/\omega_j)$, there exists a solution z_T of (H) having minimal period T. Moreover one has

$$(9) \qquad \lim_{T \to (2\pi/\omega_j)^-} \Big(\sup_{t \in [0,T]} |z_T(t)| \Big) = 0$$

An answer to the problem of the existence of periodic solutions having arbitrarily long minimal period was given by Birkhoff and Lewis in a celebrated paper of 1933 ([2], see also [7], [9] for more recent proofs). The arguments are essentially based on the use of some fixed point techniques.

Theorem 3. Let all the previous assumptions be satisfied, let H verify (3), (4), (5) and let

$$(10) \qquad \omega_h/\omega_k \notin \mathbb{Q} \quad \forall\, h,k \in \{1,\dots,N\}, \quad h \neq k$$

Then, for any prime integer number $p \geq 2$, there exists a positive number $\varepsilon = \varepsilon(p)$ such that, for any $j \in \{1,\dots,N\}$ and for any T in the interval $(2p\pi/\omega_j - \varepsilon,\ 2p\pi/\omega_j)$, there exists a solution z_T of (H) having minimal period T or T/p and the following condition is satisfied:

$$(11) \qquad \lim_{T \to (2p\pi/\omega_j)^-} \Big(\sup_{t \in [0,T]} |z_T(t)| \Big) = 0$$

3. Here we present some concepts and ideas which are basic for the proofs of Theorems 1, 2, 3.

A. The action duality principle. Let $T > 0$, $T \neq 2k\pi/\omega_j$, $k \in \mathbb{N}$, $j \in \{1,\dots,N\}$, let $\mathscr{L}_T: H^{1,\alpha}(R/T\mathbb{Z};R^{2N}) \to L^\alpha(0,T;R^{2N})$ (with $\alpha = \beta/\beta - 1$) be the operator defined as

$$\mathscr{L}_T v = J\dot{v} - Qv$$

and let us consider the legendre transform of H, that is

$$G(v) = \sup\{<v,w> - H(w): w \in R^{2N}\} \quad \forall\, v \in R^{2N}$$

Then, setting

$$F_T(v) = \int_0^T G(v) - \tfrac{1}{2}\int_0^T <\mathscr{L}_T^{-1} v, v> \qquad \forall\, v \in L^\alpha(0,T;R^{2N}) ,$$

it is easy to show that $F_T'(u_T) = 0$ iff $z_T = \mathscr{L}_T^{-1} u_T$ is T-periodic solution of (H).

B. Critical points of Mountain Pass type. One can prove that F_T satisfies the following properties:

(B_1) F_T verifies the Palais-Smale condition

(B_2) $F_T(v) > 0$ for $v = 0$, $||v||_\alpha < \rho$

(B_3) There exists some $\bar{e} \in L^\alpha$ such that $F_T(\bar{e}) < 0$

Then, by applying the well known theorem by Ambrosetti and Rabinowitz (see [1]), one obtains the existence of a critical point u_T of F_T of Mountain Pass type, that is, for all open neighbourhood $\mathcal{U}$ of u_T, the set

$$\{v \in \mathcal{U} : F_T(v) < F_T(u_T)\}$$

is neither empty nor path connected (see [4]).

Also, putting $z_T = \mathcal{L}_T^{-1} u_T$, one has

$$\text{(12)} \qquad \lim_{T \to (2k\pi/\omega_j)^-} \left(\sup_{t \in [0,T]} |z_T(t)| \right) = 0$$

C. <u>The Morse index of a critical point of</u> F_T. Let u_T be a given critical point of F_T. For any $s \in (0,T]$, the <u>Morse index</u> of the quadratic form

$$Q_s^T(v) = \langle F_T''(u_T)v, v\rangle = \int_0^T \langle G''(u_T)v, v\rangle - \int_0^T \langle \mathcal{L}_T^{-1} v, v\rangle \quad \forall v \in L^2(0,s;R^{2N})$$

is defined as the dimension of the maximal subspace of $L^2(0,s;R^{2N})$ where Q_s^T is negative definite. For $s = T$, one gets the Morse index of the point u_T, denoted by $m(u_T)$.

D. <u>The Morse index of a critical point of Mountain Pass type</u>. By a finite dimensional reduction and a generalized version of the Morse lemma due to Hofer [8], it is possible to show that the Morse index of a critical point u_T of Mountain Pass type satisfies the condition

$$\text{(13)} \qquad m(u_T) \leqq 1$$

E. <u>Conjugate points and minimal periods</u>. Let z_T be a T-periodic solution of (H) and let $s \in (0,T)$. We will say that $z_T(s)$ is <u>conjugate</u> to $z_T(0)$ if the linear space L_s of the solutions y of the system

$$J\dot{y}(t) = Qy(t) + H''(z_T(t))y(t) \qquad \forall\, t \in (0,s)$$
$$y(0) = y(s)$$

is different from {0}. The <u>multiplicity</u> of $z_T(s)$ is defined as the dimension of L_s.

Denoting by $O(z_T)$ the maximum integer number k such that T/k is a period of z_T (that is T/k is the minimal period of z_T), by taking into account that, if z_T is s-periodic, then $z_T(s)$ is conjugate to $z_T(0)$ (as $\dot{z}_T$ belongs, in such a case, to L_s), it is obvious that $O(z_T) - 1$ is less or equal to the number of conjugate points $z_T(s)$ to $z_T(0)$ when s varies in (0,T).

F. <u>The Morse index formula</u>. Let z_t be a T-periodic solution of (H), then there exists only a finite number r of points $z_T(s_i)$ which are conjugate to $z_T(0)$. Denoting by m_i the relative multiplicity, the following index formula holds:

$$(14) \qquad m(u_T) = \sum_{i=1}^{r} m_i - 2 \sum_{i=1}^{N} [T\omega_j/2\pi]$$

where $u_T = Z_T z_T$ and [x] is the integer part of x (see [5]).

5. Now we can give an outline of the proofs of Theorems 1, 2, 3. As for Theorem 1, it is sufficient to use formulas (13), (14) in order to obtain, for any $T < 2\pi/\omega_N$, a critical point u_T of Mountain Pass type whose corresponding solution $z_T = \mathscr{L}_T^{-1} u_T$ has minimal period T or T/2. Afterwards, a suitable argument based on the properties of a critical point of Mountain Pass type excludes that T/2 can be a period of z_T,

so z_T has minimal period T.

The thesis of Theorem 2 follows from (10), (12) and from an appropriate analysis of the index formula (14).

Let us give now the ideas of the proof of Theorem 3. Firstly, let us put some notations:

$$\Omega = \{2k\pi/\omega_j : k \in \mathbb{N},\ j \in \{1,\dots,N\}\} \qquad \Omega'_T = \Omega \cap (0,T] \qquad \forall\, T > 0$$

$\{Q_s^T\}_{\substack{s \in (0,T)\setminus\Omega'_T \\ T \in R_+\setminus\Omega}}$ is the family of quadratic forms associated with a fixed family of critical points u_T of Mountain Pass type.

Then, as a consequence of (12), it is possible to show that, given two consecutive elements $\bar{s}_1$, $\bar{s}_2$ in Ω, for any pair of consecutive elements s_1, s_2 in $\Omega'_{\bar{s}_1}\setminus\{0\}$ and for any $\varepsilon \in (0, s_2 - s_1)$, there exists some $\delta = \delta(\varepsilon)$ in $(0, \bar{s}_2 - \bar{s}_1)$, such that, if $T \in (\bar{s}_2 - \delta, \bar{s}_2)$, then Q_s^T is positive definite for all $s \in (s_1, s_2 - \varepsilon)$.

At this point, it is easy to show that, when T lies in a sufficiently small left neighbourhood of $2p\pi/\omega_j$, where p is a prime number, then Q_s^T is positive definite for all $s = T/n$, with $n \geq 2$, $n \neq p$, so T/n with $n \geq 2$, cannot be a period of z_T, unless $n = p$, and the proof of Theorem 3 is concluded.

REFERENCES

[1] Ambrosetti, A., Rabinowitz, P.: "Dual variational methods in critical point theory and applications", J. Funct. Anal. 14(1973) 349-381

[2] Birkhoff, G.D., Lewis, D.C.: "On the periodic motions near a given periodic motion of a dynamical system", Ann. Mat. Pura e Appl. 12 (1933), 117-133

[3] Clarke, F., Ekeland, I.: "Hamiltonian trajectories having prescribed minimal period", Comm. Pure and Appl. Math. 33 (1980), 103-116

[4] Ekeland, I., Hofer, H.: "Periodic solutions with prescribed period for convex autonomous Hamiltonian systems", preprint Ceremade n. 8421, Paris (1984)

[5] Girardi, M., Matzeu, M.:"Periodic solutions of convex Hamiltonian systems with quadratic growth at the origin and superquadratic at infinity", to appear on Ann. Mat. Pura e Appl.

[6] Girardi, M., Matzeu, M.: "Solutions of minimal period for Hamiltonian systems with a quadratic growth at the origin and superquadratic at infinity", to appear on Rend. Ist. Mat. Univ. Trieste

[7] Harris, T.C.: "Periodic solutions of arbitrarily long periods in Hamiltonian systems", J. Diff. Eq. 1 (1968), 131-141

[8] Hofer, H.: "A geometric description of the neighbourhood of a critical point given by the Mountain Pass theorem", J. London Math. Soc., to appear

[9] Moser, J.: "Proof of a generalized form of a fixed point theorem due to G.D. Birkhoff", in "Geometry and Topology", Springer Lect. Notes in Math. n. 597 (1977), 464-494

[10] Rabinowitz, P.: "Periodic solutions of Hamiltonian systems", Comm. Pure and Appl. Math., 31 (1978), 157-184.

RELATIONS BETWEEN GLOBAL INVARIANTS OF CONVEX CONTACT MANIFOLDS AND LOCAL INVARIANTS OF THEIR PERIODIC HAMILTONIAN TRAJECTORIES

H. Hofer*

Rutgers University
Hill Center for Mathematical Sciences
New Brunswick, N.J. 08903

1. INTRODUCTION AND STATEMENT OF THE MAIN RESULTS

In this paper I shall report about a recent joint paper with I. Ekeland, [6], concerning the number of periodic Hamiltonian trajectories on a prescribed energy surface. Before I give a statement and a sketch of some of the proofs of the results let me sum up what is known and explain the problem.

Assume S is compact odd-dimensional smooth manifold equipped with a closed 2-form ω. Clearly ω must be degenerate. The standard examples are compact regular energy surfaces given by Hamiltonian $H:M \to \mathbb{R}$, defined on some symplectic manifold (M,Ω), where we take $\omega := \Omega|S$.

Following essentially Weinstein [21], we define

Definition 1. (S,ω) is called to be of contact type if there exists a 1-form λ on S such that

(1) $\quad d\lambda = \omega$ and $\lambda\wedge\omega^{n-1}$ is a volume,

where $\dim S = 2n-1$ and $\omega^k = \omega\wedge\ldots\wedge\omega$ (k times). We refer to (S,ω) as

---

* Research partially sponsored by University Research Council Grant.

a contact type manifold and call λ satisfying (1) an admissible 1-form for (S,ω).

Condition (1) implies that ω is as good as it can be, namely ω has a one-dimensional kernel. Therefore ω defines a one-dimensional line-bundle $\mathscr{L}_S \to S$, where the fibre over $x \in S$ consists of all $v \in T_x S$ annihilating ω_x, that is $v \rightharpoonup \omega_x = 0$. We shall call $\mathscr{L}_S$ the characteristic distribution of (S,ω). Having a distribution one can ask for integral manifolds; an important class of integral manifolds G are those which are diffeomorphic to S^1.

Definition 2. Let (S,ω) be as described above. A periodic Hamiltonian trajectory is a submanifold G of S, diffeomorphic to S^1 such that

$$(2) \qquad TG = \mathscr{L}_S | G$$

For (2) note that $\mathscr{L}_S \subset TS$ and TG is identified with a subbundle of $TS|G$ via T (incl). The collection of Hamiltonian trajectories will be denoted by [S] or $[(S,\omega)]$.

Having defined [S] one can raise the question if [S] is in general "another example of" the empty set. Weinstein, [21], conjectured that this is not the case, namely

Conjecture 1 (Weinstein). Let (S,ω) be a compact simply connected manifold of contact type, then $[S] \neq \emptyset$.

The conjecture in its full generality is unproved till now. However for certain special cases much is known about S . Before we outline the results obtained in this paper we shall give a short survey of some of the known results.

Denote by j the complex on $\mathbb{R}^{2n}$ given by

$$j = \begin{pmatrix} 0_n & 1_n \\ -1_n & 0_n \end{pmatrix}$$

and let $<\cdot,\cdot>$ be the standard inner product on $\mathbb{R}^{2n}$. Then Ω given by

$$\Omega : = <j\cdot,\cdot>$$

is a symplectic form on $\mathbb{R}^{2n}$. We impose now the following hypoteses

($\mathcal{H}$) $S \subset \mathbb{R}^{2n}$ is a compact smooth manifold bounding a convex region. Moreover S has non-vanishing Gaussian curvature. Let $\omega := \Omega \,|S$.

A simple lemma is the following

Lemma 1. If (S, ω) satisfies ($\mathcal{H}$) then there exists a 1-form λ on S such that (1) holds. Hence, (S, ω) is of contact type. Moreover if $n \geq 2$ S is simply connected.

The following has been proved in 78

- If (S,ω) satisfies ($\mathcal{H}$) then $[S] \neq \emptyset$ (Weinstein [20], Rabinowitz [16-17], Rabinowitz proves in fact a somewhat stronger result).

Knowing that $[S] \neq \emptyset$ one can ask for its cardinality. The worst example known satisfying ($\mathcal{H}$) has $\sharp[S] = n$. Hence there is the following conjecture.

Conjecture 2. If (S,ω) satisfies ($\mathcal{H}$) then $\sharp[S] \geq n$.

Concerning Conjecture 2 the following results are known to be true.

- $\sharp[S] \geq n$ if S encloses a sphere of radius $\underline{r}$ and is enclosed by a sphere of radius $\bar{r}$ (both spheres have the same center) such that $\bar{r} < \sqrt{2}\,\underline{r}$. (Ekeland-Lasry [9], see also [1,2,12] for different proofs and extensions).

- $\#[S] = +\infty$ for a generic set of surfaces (S,ω) satisfying $(\mathscr{H})$ if $n \geq 3$. (Generic refers to a residual set for some topology on the sets S, Ekeland [5]).

- $\#[S] \geq 2$ if $n \geq 3$. (Ekeland-Lassoued, [10]).

The purpose of this paper is to study surfaces satisfying $(\mathscr{H})$ which have only a finite number of periodic Hamiltonian trajectories and to associate to such a surface S global invariants and to $G \in [S]$ local invariants and to show that these invariants cannot be independent. Before we start with this program some more definitions for the general situation.

Assume (S,ω) is a compact manifold of contact type which is simply connected. (Consequently $H^1(S;\mathbb{R}) = 0$.) Let λ denote an admissible 1-form. Since $\lambda \wedge \omega^{n-1}$ is a volume λ will be nonzero everywhere and there exists a unique vector field ξ_λ on S such that

$$(3) \qquad \xi_\lambda \rightharpoonup \lambda \equiv 1, \quad \xi_\lambda \rightharpoonup \omega \equiv 0$$

One easily verifies that if τ is another admissible 1-form then $\xi_\tau = f\,\xi_\lambda$ for some positive smooth function f on S. Hence the orientation of $\mathscr{L}_S$ induced by ξ_λ is independent of λ. This orientation induces an orientation for $G \in [S]$.

<u>Definition 3</u>. The volume of $G \in [S]$ is defined by

(4) $\mathrm{vol}(G) := \int \lambda|G$

Note that this definition is independent of the choice of λ since $H^1(S;\mathbb{R}) = 0$. Next consider the initial value problems

$$\dot{x} = \xi_\lambda(x)$$
$$x(0) = x_\circ \in G$$

for some $G \in [S]$ and $x_\circ \in G$. This differential equation has a unique solution $x_\lambda\colon \mathbb{R} \to G$ with minimal period $T_\lambda > 0$, say. Clearly

$$\mathrm{vol}(G) = \int \lambda|G = \int_0^{T_\lambda} x_\lambda^* \lambda = \int_0^{T_\lambda} 1 = T_\lambda$$

So the minimal period of x_λ does not depend on λ. Linearizing the above differential equation x_λ gives

(5) $$\dot{X}(t) = T\xi_\lambda(x_\lambda(t))X(t)$$
$$X(0) = X_\circ \in T_{x_\circ} S$$

where $x_\circ$ is given. (5) induces for $t \in \mathbb{R}$ a linear map

$$R_\lambda(t)\colon T_{x_\circ} S \to T_{x_\lambda(t)} S$$

In particular we obtain a map $R_\lambda\colon T_{x_\circ} S \to T_{x_\circ} S$ given by

(6) $$R_\lambda := R_\lambda(T_\lambda) = R_\lambda(\mathrm{vol}(G))$$

One verifies easily that if λ and τ are admissible then R_λ and R_τ are conjugate. Moreover if we change $x_\circ$ we get conjugate maps. Hence the spectrum of R_λ is an invariant for $G \in [S]$.

Definition 4. The spectrum of $G \in [S]$ denoted by $\sigma(G)$ is the set of eigenvalues $\mu_1,\dots,\mu_{2n-1}$ (each repeated according to its multiplicity) of $R_\lambda \in \mathscr{L}(T_{x_0}S)$ for some admissible λ. (Note that always $1 \in \sigma(G)$ since $R_\lambda \xi_\lambda(x_0) = \xi_\lambda(x_0)$.) We call G nondegenerate if the geometric multiplicity of $1 \in \sigma(G)$ is one. G is called strongly non-degenerate if it is nondegenerate and all eigenvalues different from 1 are not a root of unity.

We shall restrict ourselves now to hypersurfaces $S \subset \mathbb{R}^{2n}$ satisfying $(\mathscr{H})$. Translating S in $\mathbb{R}^{2n}$ we may assume that S encloses $0 \in \mathbb{R}^{2n}$. Note that the translation of $[S]$ is [translation of S].

Definition 5. We denote by $\mathscr{H}$ the collection of all smooth hypersurfaces S in $\mathbb{R}^{2n}$ enclosing $0 \in \mathbb{R}^{2n}$ and satisfying $(\mathscr{H})$. Moreover we denote by $\tilde{\mathscr{H}}$ the collection of all maps $H\colon \mathbb{R}^{2n} \to \mathbb{R}$ such that

$$(7) \qquad \begin{aligned} &H \in C(\mathbb{R}^{2n},\mathbb{R}) \cap C^\infty(\mathbb{R}^{2n}\setminus\{0\},\mathbb{R}) \\ &H(\lambda x) = \lambda^2 H(x) \quad \text{for all} \quad x \in \mathbb{R}^{2n},\ \lambda \geq 0 \\ &H''(x) \geq \alpha_x \mathrm{Id} \quad \text{for } x \in \mathbb{R}^{2n}\setminus\{0\},\ \alpha_x > 0 \end{aligned}$$

Here $H''(x) \in \mathscr{L}(\mathbb{R}^{2n})$ denote the linearization of the gradient $H'(x)$ of H at $x \in \mathbb{R}^{2n}$. The following result is obvious.

Lemma 2. There is a natural bijection

$$\mathscr{H} \to \tilde{\mathscr{H}}$$

associating to S the unique function H_S in $\mathscr{H}$ satisfying $H_S^{-1}(1) = S$.

It is clear that there is an obvious bijection between $C^\infty(S^{2n-1};\mathbb{R})$ and the set $\{H \in C(\mathbb{R}^{2n};\mathbb{R}) \cap C^\infty(\mathbb{R}^{2n}\setminus\{0\};\mathbb{R}) \mid H(\lambda x) = \lambda^2 H(x)$ for $x \in \mathbb{R}^{2n}$ and $\lambda \geq 0$. Since $C^\infty(S^{2n-1};\mathbb{R})$ has a topology coming from a complete metric we obtain induced topologies for $\tilde{\mathscr{H}}$ and $\mathscr{H}$ which we call the strong topologies. It is quite easy to show that $\tilde{\mathscr{H}}$ and $\mathscr{H}$ have the

Baire property with respect to the strong topology, that is a countable intersection of open dense sets is dense.

Let $H \in \tilde{\mathscr{H}}$. The Fenchel conjugate of H denoted by H* is the function $\mathbb{R}^{2n} \to \mathbb{R}$ defined by

(8) $$H^*(y) = \max_{x \in \mathbb{R}^{2n}} (\langle x,y \rangle - H(x))$$

It is well known that $H^* \in \tilde{\mathscr{H}}$. Note that $*: \tilde{\mathscr{H}} \to \tilde{\mathscr{H}}$ and the induced map, also denoted by *, $\mathscr{H} \to \mathscr{H}$ are continuous for the strong topologies.

It is clear that $H \in \tilde{\mathscr{H}}$ has actually a globally Lipschitz continuous gradient H'.

We introduce now a Hilbert space E by

$$E = \{x: S^1 = \mathbb{R}/\mathbb{Z} \to \mathbb{R}^{2n} \mid x \text{ is absolutely continuous and has a square integrable derivative. Further } \int_0^1 x(t)dt = 0 .$$

As an inner product we take

$$(x,y) = \int_0^1 \langle \dot{x}(t), \dot{y}(t) \rangle \, dt.$$

Now we introduce for $S \in \mathscr{H}$ a $C^{1,1}$-Hilbert manifold M_S by

$$M_S = \{x \in E \mid \int_0^1 H_S^*(-j\dot{x}(t))dt = 1\} .$$

We have a natural S^1 action on E denoted by

$$S^1 \times E \to E; \quad (a,x) \to a*x$$

where S^1 acts by phase shift. The induced action on M_S turns M_S into a S^1-space. We define a smooth map $A: E \to \mathbb{R}$ by

(9) $$A(x) = \int_0^1 \frac{1}{2} \langle j\dot{x}, x\rangle \, dt .$$

Clearly A is S^1-invariant. The restriction of A to M_S will be denoted by A_S. Denote by $(E_\circ, p_\circ, B_\circ)$ the principal S^1-bundle (see Husemoller [14])

$$E_\circ = \bigcup_{n \in \mathbb{N}} S^{2n+1} =: S^\infty$$

$$B_\circ = \bigcup_{n \in \mathbb{N}} \mathbb{C}P^n =: \mathbb{C}P^\infty$$

We obtain a covariant functor F from the category of paracompact S^1-spaces to the topological category by

$$X \to (X \times E_\circ)/S^1$$

where S^1 acts freely on $X \times E_\circ$ in the obvious way and $(X \times E_\circ)/S^1$ is the quotient by this action. Denote by $\overline{H}$ Alexander-Spanier Cohomology with coefficients in $\mathbb{Q}$. Then the composite functor $h := \overline{H} \circ F$ is an S^1-cohomology theory on the category of paracompact S^1-spaces.

Given $S, R \in \mathscr{H}$ there is a unique S^1-homeomorphism $\phi_{SR}: M_R \to M_S$ of the form

$$\phi_{SR}(x) = f(x)x$$

for some positive function f. Since $A_S \circ \phi_{SR}(x) = f(x)^2 A_R(x)$ we see that ϕ_{SR} preserves the sign of A. For $S \in \mathscr{H}$ we denote by $\hat{M}_S$ the set

$$\hat{M}_S = \{x \in M_S \mid A(x) < 0\} .$$

By the preceding discussion ϕ_{SR} induces a homeomorphism again denoted by ϕ_{SR} from $\hat{M}_R$ to $\hat{M}_S$. A crucial topological result is the following

Theorem 1. For each given $S \in \mathscr{H}$ there exists a cohomology class $a_S \in h^2(\hat{M}_S)$ such that

$$h(\hat{M}_S) = \mathbb{Q}\, a_S$$

and $\phi^*_{SR}(a_S) = a_R$ for all $R \in \mathscr{H}$.

We can define now for $S \in \mathscr{H}$ a monotonic map $(-\infty, 0) \to \mathbb{N}$, $\mathbb{N} = \{0,1,2,\dots\}$ as follows

Definition 6. For $S \in \mathscr{H}$ a map $\alpha_S : (-\infty, 0) \to \mathbb{N}$ is defined as follows

(10) $$\alpha_S(d) = \inf\{k \in \mathbb{N} \mid i^*(a_S^k) = 0\}$$

where $i : A_S^{-1}((-\infty, d]) \to \hat{M}_S$ is the inclusion map.

One has of course to verify that α_S is well defined. From the commutative diagram for $c < d$

$$\begin{array}{ccc} A_S^{-1}((-\infty,d]) & & \hat{M}_S \\ \text{incl} & \text{incl} & \\ A_S^{-1}((-\infty,c]) & & \text{incl} \end{array}$$

it follows at once that α_S is nondecreasing. It can be verified using the strong continuity of Alexander-Spanier Cohomology that

(11) $$\lim_{d \downarrow d_\circ} \alpha_S(d) = \alpha_S(d_\circ)$$

Assume now $R \in \mathscr{H}$ such that R encloses S. Then

$$H_R \le H_S \iff H^*_R \ge H^*_S$$

Hence we obtain the following commutative diagram for some $d \in (-\infty,0)$, where all maps are induced by the ϕ's or by inclusions

$$\begin{array}{ccc} A_R^{-1}(-\infty,d] & \Phi & A_S^{-1}(-\infty,d] \\ i & & j \\ M_R & \phi_{SR} & M_S \end{array}$$

Assume $\alpha_S(d) = k$. Then $j^*(a_S^k) = 0$. This implies

$$i^*(a_R^k) = i^*\phi_{SR}^*(a_S^k)$$

$$= \phi^* j^*(a_S^k)$$

$$= 0$$

Hence for all $\ell \geq k$ in $\mathbb{N}$ we infer $i^*(a^\ell) = i^*(a)^\ell = i^*(a)^k \cup i^*(a)^{\ell-k} =$ $= 0$. So

$$\alpha_R(d) \leq \alpha_S(d) .$$

So we have proved

<u>Lemma 3</u>. If $S,R \in \mathscr{H}$ and R encloses S then

$$\alpha_R \leq \alpha_S$$

We have moreover the following

<u>Lemma 4</u>. Let $S \in \mathscr{H}$, then

$$(12) \qquad 0 < \liminf_{d \uparrow 0} |d| \alpha_S(d) \leq \limsup_{d \uparrow 0} |d| \alpha_S(d) < +\infty .$$

In view of Lemma 3 it is for example enough to show that the statement is correct for a sphere in $\mathscr{H}$. One can show that for a sphere the limit of $|d|\alpha_S(d)$ as $d \to 0$ exists, see [6]. In fact the following is true

Lemma 5. Let S H be a sphere of radius $r > 0$. Then

(13) $$\lim_{d \uparrow 0} |d|\alpha_S(d) = \frac{n}{\pi r^2}$$

In view of the previous results we can introduce two global invariants for $S \in \mathscr{H}$.

Definition 7 . Let $S \in \mathscr{H}$

(14) $$I^S = \limsup_{d \uparrow 0} |d|\alpha_S(d)$$

$$I_S = \liminf_{d \uparrow 0} |d|\alpha_S(d) .$$

We call I_S the lower and I^S the upper index of S.

Remark. It is an open problem if there exists $S \in \mathscr{H}$ with $I_S < I^S$. In view of the next results the existence of such a surface S would be extremely interesting.

Define a metric ρ on H by

$$\rho(R,S) = \max_{x \in R} \inf_{y \in S} |x-y| + \max_{y \in S} \inf_{x \in R} |x-y| .$$

This is of course the Hausdorff metric. It is clear that the topology induced by ρ on $\mathscr{H}$ is coarser than the strong topology. If $S \in \mathscr{H}$ and $b > 0$ we denote by bS the surface associated to $b^{-2}H_S$. One can verify easily that

(15) $$\alpha_{bS}(d) = \alpha_S(b^2d).$$

Hence

(16) $$I^{bS} = b^{-2}I^S$$
$$I_{bS} = b^{-2}I_S$$

This implies of course the following

Lemma 6. The maps $I^*, I_*: (\mathcal{H},\rho) \to \mathbb{R}$ are continuous.

We come now to the first main result

Theorem 2. If $\sharp[S] < \infty$ then

(17) $$I_S = I^S$$

Coming back to the earlier remark it follows from Theorem 2 that if there is a $S_\circ \in \mathscr{H}$ with $I_{S_\circ} < I^{S_\circ}$ then there is a $\delta = \delta(I_{S_\circ}, I^{S_\circ}) > 0$ such that $\sharp[S] = \infty$ for all $S \in \mathscr{H}$ with $\rho(S,S_\circ) < \delta$.

Next we introduce local invariants for a $G \in [S]$ and relate them to I_S and I^S. Fix $S \in \mathscr{H}$ and $G \in [S]$. One easily verifies that for the admissible 1-form $\lambda = \theta|S$ where θ is the 1-form on $\mathbb{R}^{2n}$ given by
$$\theta(x,h) = \frac{1}{2} \langle jx,h \rangle$$
we have
$$\xi_\lambda(x) = jH'(x), \ x \in S$$
where H' is the gradient of $H: \mathbb{R}^{2n} \to \mathbb{R}$. Put $x(t) := x_\lambda(t)$ with $x(0) = x \in S$ and consider the fundamental solution of the linear differential equation in $\mathbb{R}^{2n}$ given by

$$\dot{R}(t) = jH''(x(t))R(t)$$

$$R(0) = Id_{\mathbb{R}^{2n}}$$

Denote by R^* the adjoint of R. Define $B: \mathbb{R} \to Sp(n, \mathbb{R})$ by

$$B(t) = (R(t)R(t)^*)^{-\frac{1}{2}} R(t)$$

Then $|B(t)z| = |z|$ for all $t \in \mathbb{R}$ and $z \in \mathbb{R}^{2n}$ and $B(t)$ commutes with j. j defines a complex multiplication on $\mathbb{R}^{2n}$ by $iz := jz$. Denoting by $\det_j$ a nonzero corresponding complex determinant function we find a unique continuous map $\Delta_G: \mathbb{R} \to \mathbb{R}$ characterized by

(18) $$\Delta_G(0) = 0$$

$$\det_j \circ (\underbrace{B(t) \ldots B(t)}_{n\text{-times}}) = e^{2\pi i \Delta_G(t)} \det_j$$

Δ_G means in some sense the (symplectic or contact) torsion around G. Therefore we define

<u>Definition 8</u>. The contact torsion of $G \in [S]$ is the real number

(19) $$\begin{aligned} \gamma_G &:= \Delta_G(\mathrm{vol}(G)) \\ &= \Delta_G(T_\lambda) \\ &= \Delta_G \ (\text{minimal period of } x) \end{aligned}$$

The contact torsion of $G \in [S]$ per volume unit is the real number

(20) $$\bar{\gamma}_G := \gamma_G / \mathrm{vol}(G) .$$

The next main result is the following (see [6])

Theorem 3. For $S \in \mathscr{H}$ we have

(i) If $n \geq 2$ then $\gamma_G > 1$ for all $G \in [S]$.

(ii) If $n > 1$ and $[S]$ is finite then there exist two different $G_1, G_2 \in [S]$ such that

$$\bar{\gamma}_{G_1} = \bar{\gamma}_{G_2} = I_S = I^S$$

(iii) Under the hypothesis of (ii) with $I(S) := I^S = I_S$ we have the even stronger result

$$\sum_{G \in [S], \bar{\gamma}_G = I(S)} \gamma_G^{-1} \geq 1$$

or with other words

$$\sum_{G \in [S], \bar{\gamma}_G = I(S)} \mathrm{vol}(G)^{-1} \geq I(S)$$

The relations given in (iii) are sharp. For examples if S is the surface given by

$$H(q \ldots q_n, p \ldots p_n) = \frac{1}{2} \sum_{i=1}^{n} \alpha_i (q_i^2 + p_i^2) \tag{20}$$

where the $\alpha_i > 0$ are independent over $\mathbb{Z}$ then

$$\#[S] = n$$

$$I(S) = \frac{1}{2\pi} \sum_{i=1}^{n} \alpha_i$$

$$\bar{\gamma}(G_j) = \frac{1}{2\pi} \sum_{i=1}^{n} \alpha_i \quad \text{for } j = 1 \ldots n$$

$$\bar{\gamma}(G_j) = \sum_{i=1}^{n} \frac{\alpha i}{\alpha j}$$

$$\mathrm{vol}(G_j) = \frac{2\pi}{\alpha_j}$$

Then $\sum_{i=1}^{n} \mathrm{vol}(G_j)^{-1} = \sum_{j=1}^{n} \frac{\alpha_j}{2\pi} = I(S)$. One may conjecture the following sharpening of Conjecture 2.

Conjecture 3. If $S \in \mathscr{H}$ and [S] is finite then there exist $G, \ldots, G_n$ [S] with $\overline{\gamma}(G_i) = I(S)$.

Next we have a look at the generic situation.

Definition 9. $\mathscr{H}_1$ denotes the subspace of $\mathscr{H}$ consisting of all S such that all $G \in [S]$ are strongly nondegenerate.

Using a version of Thom's transversality theorem we arrive at

Lemma 7. $\mathscr{H}_1$ is a residual subset of $\mathscr{H}$ for the strong topology, in particular $\mathscr{H}_1$ is dense.

For example the surface S associated to H defined in (20) belongs to $\mathscr{H}_1$.

So $\mathscr{H}_1$ contains surfaces S with $\sharp[S] < \infty$.

The main result for surfaces in $\mathscr{H}_1$ is

Theorem 4. Assume $S \in \mathscr{H}_1$ and [S] is finite, then there exist numbers $\varepsilon_G \in \{-1, -\frac{1}{2}, \frac{1}{2}, 1\}$ such that

$$\sum_{G \in [S]} \varepsilon_G \gamma_G^{-1} = 1$$

If moreover all $\varepsilon_G = 1$ then $\overline{\gamma}_G = I(S)$ for all $G \in [S]$. (In fact there is a precise way to decide the value of ε_G.)

It would be interesting to prove Conjecture 3 under the assumption $S \in \mathscr{H}_1$.

Therefore

Conjecture 4 (weak Conjecture 3) — If $S \in \mathscr{H}_1$ and [S] is finite then there exist $G \ldots G_n \in [S]$ with

$$\bar{\gamma}_{G_i} = I(S) \quad \text{for } i = 1 \ldots n$$

Next we pass to another residual set.

Definition 10. The set of all $S \in \mathscr{H}$ such that $\bar{\gamma}: [S] \to \mathbb{R}$ is injective is denoted by $\mathscr{H}_2$.

Using Thom's transversality theorem again we have

Lemma 8 $\mathscr{H}_2$ is a residual subset of $\mathscr{H}$, in particular it is dense.

Lemma 8 has the following nice corollary.

Theorem 5. For all $S \in \mathscr{H}_2$ we have $\#[S] = +\infty$ if $n \geq 2$.

The proof is quite simple. If [S] is finite then there exist G_1, G_2 in [S] for which by Theorem 3 (ii) $\bar{\gamma}_{G_1} = \bar{\gamma}_{G_2}$ contradicting the fact that $\bar{\gamma}$ is injective for $S \in \mathscr{H}_2$.

Finally we give a concrete application. Consider the two-dimensional system

$$(21) \qquad \begin{aligned} \ddot{q}_1 + q_1 &= 3q_1^2 q_2^2 \\ \ddot{q}_2 + q_2 &= 2q_1^2 q_2 \end{aligned}$$

It can be written as $\ddot{q} + V'(q) = 0$ with potential V given by

$$V(q_1,q_2) = \frac{1}{2} q_1^2 + \frac{1}{2} q_2^2 - 3q_1^3 q_2^2.$$

The Hamiltonian is $H(q_1,q_2,p_1,p_2) = \frac{1}{2} p_1^2 + \frac{1}{2} p_2^2 + V(q_1,q_2)$. For h 0 small the surface S_h close to 0 with energy h will be strongly convex.

There are two obvious periodic solutions of (21) namely

$$q_1(t) = 0, \; q_2(t) = \sqrt{2h} \cos t$$
$$q_1(t) = \sqrt{2h} \cos t, \; q_2(t) = 0$$

One can show that the corresponding Hamiltonian trajectories G_1 and G_2 in $[S_h]$ have different values $\overline{\gamma}_{G_1}$ and $\overline{\gamma}_{G_2}$ provided $h < \frac{1}{4}$ (see [6]). Hence by Theorem 3 there must be a third one $G_3 \in [S_h]$ if $h < \frac{1}{4}$ and S_h is strongly convex.

2. SKETCH OF THE PROOFS OF THE MAIN RESULTS

If $x \in \hat{M}_S$ is a critical point of A_S then we find some number $\delta < 0$ such that

$$(1) \qquad \int \langle j\dot{x}, h \rangle \, dt = \delta \int H^{*\prime}(-j\dot{x}), -j\dot{h} \rangle \, dt$$

for all $h \in E$. Hence taking in (1) $h = x$

$$(2) \qquad 0 > A_S(x) = \delta$$

Moreover for some constant $c \in \mathbb{R}^{2n}$ we have

$$(3) \qquad |\delta|^{-1}(x + c) = H^{*\prime}(-j\dot{x})$$

or equivalently

$$|\delta|^{-1} H'(x + c) = -j(x + c)^{\bullet}$$

So after multiplying with $x + c$ and integrating over $[0,1]$ we find

$$|\delta|^{-1} H(x(t)+c) = -A_S(x) .$$

So

(4) $$H(x(t)+c) = A_S(x)^2$$

So $y\colon \mathbb{R} \to \mathbb{R}^{2n}$ defined by

$$y(t) = A_S(x)^{-1}(x(t)+c)$$

parametrizes a $G \in [S]$, where $S = S_H$. If S^1 acts freely on x or with other words if x has minimal period 1 we have therefore for $G = G_y$

(5)
$$\begin{aligned} \mathrm{vol}(G) &= \left| \int_0^1 y^* \lambda \, dt \right| \\ &= A_S(x)^{-2} \left| \int_0^1 x^* \lambda \, dt \right| \\ &= |A_S(x)^{-1}| \\ &= |\delta|^{-1} \end{aligned}$$

Now if $x \in \hat{M}_S$ is a critical point of A_S so is for $k \in \mathbb{N}^* = \{1,2,\ldots\}$ $x_k \in \hat{M}_S$ given by

(6) $$x_k(t) = \frac{1}{k} x(kt)$$

Clearly $A_S(x_k) = \frac{1}{k} A_S(x)$. The functional A_S is unfortunately only C^1 but not C^2. However we can formally linearize at a critical point $x \in \hat{M}_S$ of A_S. The Morse index will be denoted by $m^-(x)$ and is known to be finite, see for example [5, 7, 8]. As we have seen infinitely many critical points, namely $x = x_1, x_2, x_3$, give the same Hamiltonian trajectory $G \in [S]$ after some reparametrisation and normalization. The sequence of Morse indices $m^-(x_k)$ defines a sequence (i_k),

$i_k = i_k(G)$. It has been shown that $\lim_{k\to+\infty} i_k/k = \bar{i}$ exists, see [5]. Using also the results by Conley and Zehnder, [4] we infer that

(7) $$\bar{i}(G) = 2\,\gamma_G$$

Denote by $m_k^\circ = m_k^\circ(G)$ the nullity of the linearization of A_S at x_k. By a hard variation of a result by Viterbo [18], (we have not a C^2-functional) we can show the following

Proposition 1. If $d \in (-\infty, 0)$ is a point of discontinuity of α_S then there exists a critical point x of A_S on level D. If moreover [S] is finite then the following relation holds for the Morse index and the nullity of a suitable critical point y on level d

(8) $$2\alpha_S(d) - 1 - m^\circ(y) \leq m^-(y) \leq 2\alpha_S(d) - 2$$

Moreover if [S] is finite α_S can jump by exactly 1 at a point of discontinuity.

Now assume [S] is finite. So let $[S] = \{G_1 \ldots G_L\}$ and x_k, j = = 1...L, $k \in \mathbb{N}^*$ the Towers of critical points corresponding to the G_j.

Definition 1. We say $G \in [S]$ is essential if the associated sequence (x_k) of critical points of A_S satisfies the following: there exists a monotonic injection $\phi : \mathbb{N}^* \to \mathbb{N}^*$ such that for some constant $c > 0$

(9) $$|m^-(x_{\phi(k)}) - 2\alpha_S(A_S(x_{\phi(k)}))| \leq c$$

for all $k \in \mathbb{N}^*$.

It is a consequence of Proposition 1 that there exists at least one essential G if [S] is finite.

As a consequence of results in [5] and [4] we obtain from (7)

Lemma 1. Given $G \in [S]$ there exists a constant $c > 0$ such that

(10) $$|i_k(G) - 2\gamma_G k| \leq c$$

for all k $\mathbb{N}$, and

(11) $$\gamma_G = \frac{1}{2}\bar{i} > 1 \quad \text{if } n \geq 2.$$

Lemma 2. Assume $G \in [S]$ is essential. Then

(12) $$I_S \leq \bar{\gamma}_G \leq I^S$$

Proof. We have using (9) and (10) for some injective monotonic map $\phi : \mathbb{N}^* \to \mathbb{N}^*$

(13) $$|i_{\phi(k)}(G) - 2\gamma_G \phi(k)| \leq c$$
$$|i_{\phi(k)}(G) - 2\alpha_S(A_S(x_{\phi(k)}))| \leq c$$

Hence

(14) $$|\alpha_S(A_S(x_{\phi(k)})) - \gamma_G \phi(k)| \leq c.$$

Since $A_S(x_{\phi(k)}) = \frac{1}{\phi(k)} A_S(x)$ we infer from (13)

(15) $$|\alpha_S(\frac{1}{\phi(k)} A_S(x)) - \gamma_G \phi(k)| \leq c .$$

So multiplying (14) by $|\frac{1}{\phi(k)} A_S(x)|$ we infer

$$(16) \qquad \alpha_S\left(\frac{1}{\phi(k)} A_S(x)\right)\left|\frac{1}{\phi(k)} A_S(x)\right| - c_1 \frac{1}{\phi(k)}$$

$$\leq \gamma_G(A) A_S(x)$$

$$\leq \alpha_S\left(\frac{1}{\phi(k)} A_S(x)\right)\left|\frac{1}{\phi(k)} A_S(x)\right| + c_1 \frac{1}{\phi(k)}$$

for all $k \in \mathbb{N}^*$. Since we have seen already that $A_S(x) = \mathrm{vol}(G)^{-1}$ we infer talking the limit that (12) holds.

Proof of Theorem 2 and Theorem 3. Assume S is finite and let $\{G_1 \ldots G_N\}$ be the essential trajectories so that $i \leq j$ implies $\overline{\gamma}_{G_i} \leq \overline{\gamma}_{G_j}$. We show first that

$$(17) \qquad I_S = I^S = \overline{\gamma}_{G_i} \quad \text{for } i = 1 \ldots N .$$

Choose a number $c \in \mathbb{N}^*$ such that for every $G_i \in [S]_{\text{essential}}$ (13) holds, for some injective monotonic map $\phi_i : \mathbb{N}^* \to \mathbb{N}^*$. Assume that ϕ_i is maximal with this property, that is there is no $\phi \leq \phi_i$ having the same property. Pick $K \in \{1, .., N\}$ such that

$$\overline{\gamma} := \overline{\gamma}_G = \ldots = \overline{\gamma}_{G_K} < \overline{\gamma}_{G_{K+1}}$$

If $K = N$ it only remains to establish that $I_S = I^S = \overline{\gamma}$. So assume first $K < N$. Define $\tilde{\phi}_i(k) = \alpha_S(A(x^i_{\phi_i(k)}))$. If now $\#(\mathrm{im}(\tilde{\phi}_{K+1}) \cap \mathrm{im}(\tilde{\phi}_i)) = \infty$ for some $i \in \{1, \ldots, K\}$ then recalling the proof of Lemma 2 we must have

$$\overline{\gamma}_{G_i} = \overline{\gamma}_{G_{K+1}}$$

which is a contradiction. So we may assume that $\mathrm{im}(\tilde{\phi}_{K+1}) \cap \mathrm{im}(\tilde{\phi}_i)$ is is a finite set for all $i \in \{1,\dots,K\}$. Since $[S]_{essential}$ is finite we find in this case $j \in \{K+1,\dots,N\}$ and $G_i \in \{1,\dots,K\}$ and injective monotonic maps $\psi_j, \psi_i : \mathbb{N}^* \to \mathbb{N}^*$ such that

$$\alpha_S(A_S(x^j_{\psi_j(k)})) = \alpha_S(A_S(x^i_{\psi_i(k)}) + 1$$

$$|\gamma_{G_j} \psi_j(k) - \alpha(A_S(x^i_{\psi_i(k)}))| \leq c$$

Since α_S is monotonic we infer

$$0 > A_S(x^i_{\psi_j(k)}) \geq A_S(x^i_{\psi_i(k)}) .$$

Now

$$(\alpha_S(\frac{1}{\psi_i(k)} A_S(x^i)) + 1)|\frac{1}{\psi_i(k)} A_S(x^i)|$$

$$\geq (\alpha_S(\frac{1}{\psi_i(k)} A_S(x^i)) + 1)|\frac{1}{\psi_i(k)} A_S(x^j)|$$

$$\geq \alpha_S(\frac{1}{\psi_j(k)} A_S(x^j))|\frac{1}{\psi_j(k)} A_S(x^j)|$$

Taking the limit gives

$$\overline{\gamma} = \overline{\gamma}_{G_i} \geq \overline{\gamma}_{G_j} > \overline{\gamma}.$$

This contradiction proves that $\overline{\gamma}: [S]_{essential} \to \mathbb{R}$ is a constant function. Next we show that $I_S = I^S = \overline{\gamma}$. We know already that

$$I_S \leq \overline{\gamma}_G = \dots = \overline{\gamma}_{G_N} \leq I^S$$

where $[S]_{essential} = \{G_1,\dots,G_N\}$.

Arguing indirectly we may assume for example that $I_S < \overline{\gamma}_G$. Then we find a monotonic sequence (d_k) in $(-\infty, 0)$ converging to 0 such that

$$I_S = \lim_{k \to +\infty} \alpha_S(d_k)|d_k|.$$

Let d_k^+ be the next jump point on the right of d_k. Then

$$I_S \geq \limsup_{k \to +\infty} (\alpha_S(d_k^+)-1)|d_k^+|$$

$$= \limsup_{k \to +\infty} \alpha_S(d_k^+)|d_k^+|$$

Eventually taking a subsequence of the k's we may assume that the right hand side has a subsequence converging to $\overline{\gamma}$. Hence

$$I_S \geq \overline{\gamma}.$$

Similarly $I^S \leq \overline{\gamma}$. So we have

$$I_S = \overline{\gamma}_{G_1} \cdots = \overline{\gamma}_{G_K} = I^S$$

Up to know we have proved Theorem 2. Next we complete the proof of Theorem 3. We don't prove that $\gamma_G > 1$ if $n \geq 2$ but shall restrict ourselves to part (iii), which of course implies (ii). We know already that $\overline{\gamma}([S]_{essential}) = I(S)$. Let as before $[S]_{essential} = \{G_2,\dots,G_N\}$. We find $k_o \in \mathbb{N}^*$ such that for every $k \in \mathbb{N}^*$, $k \geq k_o$, there exists $m(k) \in \mathbb{N}^*$ and $j(k) \in \{1,\dots,N[$ such that in view of Proposition 1

$$\alpha_S(A_S(x_{m(k)}^{j(k)})) = k \tag{17}$$

Define $K^j(d)$ to be the greatest number m such that $A_S(x_m^j) \leq d$.

We have

$$(18)\qquad K^j(d) = \text{Integer part of } \left|\frac{A_S(x^j)}{d}\right|$$

$$= \text{Integer part of } \left|\frac{1}{d\ \mathrm{vol}(G_j)}\right|$$

$$= \text{Integer part of } \left|\frac{\bar{\gamma}_{G_j}}{\gamma_{G_j}}\ \frac{1}{d}\right|$$

Moreover obviously

$$(19)\qquad \alpha_S(A_S(x^{j(k)}_{m(k)}))-k_\circ \leq \sum_{j=1}^{N} K^j(A_S(x^{j(k)}_{m(k)}))\ .$$

Combining (18) and (19) gives using (17) and writing $c_k = A_S(x^{j(k)}_{m(k)})$

$$(20)\qquad k - k_\circ \leq \sum_{j=1}^{N} \frac{\bar{\gamma}_{G_j}}{\gamma_{G_j}} \frac{1}{|c_k|} + c.$$

Multiplying by $|c_k|$ and taking the limit

$$I(S) \leq \sum_{j=1}^{N} \frac{\bar{\gamma}_{G_j}}{\gamma_{G_j}}$$

$$= \sum_{j=1}^{N} I(S)\ \gamma_{G_j}^{-1}$$

Therefore if [S] is finite

$$(21)\qquad \sum_{G\in[S]} \gamma_G^{-1} \geq 1$$

$$\bar{\gamma}_G = I(S)$$

The generic results are based on Lemma 7 and Lemma 8 which can be proved following the methods employed in [5] or [7,8]. Theorem 4 follows from equivariant Morse theory. In fact one can establish if $[S]$ is finite and $S \in \mathscr{H}_1$ (with $[S] = \{G_1, \ldots, G_N\}$) and for example if all unstable bundles are orientable

$$\sum_{j=1}^{N} \sum_{k=1}^{\infty} t^{i_k(G_j)} = \sum_{i=1}^{\infty} t^{2i} + (1+t)Q(t).$$

From this we conclude for all $K \in \mathbb{N}^*$

$$\sum_{j=1}^{N} \sum_{i_k(G_j) \leq 2K} t^{i_k(G_j)} = \sum_{i=1}^{K} t^{2i} + (1+t)Q_K(t)$$

Evaluating at -1 gives

$$\text{(22)} \qquad \sum_{j=1}^{N} \sum_{j_k(G_j) \leq 2K} (-1)^{i_k(G_j)} = K$$

Since $\sum_{j_k(G_j) \leq 2K} (-1)^{i_k(G_j)}$ can be only either

$$\text{(23)} \qquad \#\{k \mid i_k^m \leq 2K\}$$

or

$$\text{(24)} \qquad -\#\{k \mid i_k^m \leq 2K\}$$

which of course requires some proof we find dividing (22) by K and taking the limit

$$\Sigma\, \varepsilon_G \gamma_G^{-1} = 1$$

where we used the relation

$$\lim \frac{i_k(G)}{k} = 2\gamma_G .$$

If the unstable bundles are not all orientable the argument becomes more delicate.

3. REFERENCES

[1] Ambrosetti, A., Mancini, G.:"On a theorem by Ekeland and Lasry concerning the number of periodic Hamiltonian trajectories", J. Diff. Eq. 43 (1981), p. 1-6

[2] Berestycki, H., Larsy, J.M., Mancini, G., Ruf, B.: "Existence of multiple periodic orbits on starshaped Hamiltonian surfaces", Comm. Pure and Appl. Math. Vol. 38 n. 3 (1985), p. 253-290

[3] Bott, R.: "On the iteration of closed geodesic and Sturm intersection theory", Comm. Pure and Appl. Math. 9 (1985), p. 176-206

[4] Conley, C., Zehnder, E.: "Morse-type index theory for flows and periodic solutions for Hamiltonian equations", Comm. Pure and Appl. Math., Vol. 27 (1984), p. 211-253

[5] Ekeland, I.: "Une théorie de Morse pour les systémes hamiltoniens convexes" Ann IHP "Analyse non linéaire" 1 (1984), p. 19-78

[6] Ekeland, I., Hofer, H.: "On Hamiltonian flows with finitely many closed trajectories", in preparation

[7] Ekeland, I., Hofer, H.: "Subharmonics for convex nonautonomous Hamiltonian systems" to appear Comm. Pure and Appl. Math.

[8] Ekeland, I., Hofer, H.: "Periodic solutions with prescribed minimal period for convex autonomous Hamiltonian systems", Inv. Math. 81 (1985), p. 155-188

[9] Ekeland, I., Lasry, J.M.: "On the number of periodic trajectories for a Hamiltonian flow on a convex energy surface", Ann. of Math. (2) 112 (1980), p. 183-319

[10] Ekeland, I., Lassoued,L.: "Multiplicité des trajectoires fermées de systemes Hamiltonian convexes" (to appear)

[11] Fadell, E., Rabinowitz, P.: "Generalized cohomological index theories for Lie group actions with an application to bifurcation questions for Hamiltonian systems", Inv. Math. 45 (1978), p. 139-174

[12] Hofer, H.: "A new proof for a result of Ekeland and Lasry concerning the number of periodic Hamiltonian trajectories on a prescribed energy surface", Bull. UMI (6) I-B (1982), p. 931-942

[13] Hofer, H.: "The topological degree at a critical point of mountain pass type", Proceedings AMS Summer Institute on Nonlinear Functional Analysis 1983

[14] Husemoller, D.: "Fibre Bundles", Graduate Texts on Mathematics 20, Springer

[15] Seifert, H.: "Periodische Bewegungen mechanischer Systeme", Math. Z. 51 (1948), p. 197-216

[16] Rabinowitz, P.: "Periodic Solutions of Hamiltonian Systems",Comm. Pure and Appl. Math. 31 (1978), p. 157-184

[17] Rabinowitz, P.: " Periodic solutions of a Hamiltonian system on a prescribed energy surface", J. Diff. Eq. 33 (1979), p. 336-352

[18] Viterbo, C.: "Indice des points critiques obtenue par minimax", to appear

[19] Weinstein, A.: "Normal modes for nonlinear Hamiltonian systems", Inv. Math. 20 (1973), p. 47-57

[20] Weinstein, A.: "Periodic orbits for convex Hamiltonian systems", Ann. of Math. 108 (1978), p. 507-518

[21] Weinstein, A.: "On the hypotheses of Rabinowitz's periodic orbit theorem", J. Diff. Equations 33 (1979), p. 353-358

[22] Yakubovich, V., Starzhinskii, V.: "Linear differential equations with periodic coefficients", Halstedt Press, Wiley 1980.

SUBHARMONICS WITH PRESCRIBED MINIMAL PERIOD FOR HAMINTONIAN SYSTEMS

R. Michaelek - G. Tarantello

Courant Institute of Mathematical Sciences
251 Mercer Street, New York, N.Y. 10012
U.S.A.

Consider the following Hamiltonian system:

$$(1) \qquad J\dot{z}(t) = H_z(z(t),t)$$

where $z = (z_1,\ldots,z_{2N}) \in \mathbb{R}^{2N}$; $H \in C^1(\mathbb{R}^{2N} \times \mathbb{R};\ \mathbb{R})$; J is the usual symplectic matrix

$$J = \begin{bmatrix} 0 & I_N \\ -I_N & 0 \end{bmatrix} \qquad I_N = \text{identity matrix in } \mathbb{R}^N$$

and $\dot{z}(t) = \frac{d}{dt} z(t)$, $H_z(z,t) = (\frac{\partial}{\partial z_1} H(z,t),\ldots,\frac{\partial}{\partial z_{2N}} H(z,t)$.

Assume that H is periodic in the t-variable, namely

$$H(z,t+T) = H(z,t) \qquad \forall\, z \in \mathbb{R}^{2N},\ \ \forall\, t \in \mathbb{R}$$

for some $T > 0$.

Without loss of generality we may take $T = 2\pi$.

We are interested in finding <u>subharmonics</u> for the Hamiltonian system (1), that is the periodic solutions with period an integral multiple of 2π.

Namely we seek solutions for the following problem:

$$(1)_p \quad \begin{cases} J\dot{z} = H_z(z,t) \\ z(0) = z(2\pi p) \end{cases}$$

$p > 1$ integer.

Moreover we shall require that, for the solution of $(1)_p$ the period $2\pi p$ is minimal.

We make the following assumptions on H:

(1) $H(\cdot,t)$ is convex for any fixed $t \in [0,2\pi]$

(2) There exist constants a_1, $a_2 > 0$ and $1 < \beta < 2$ such that:

$$\frac{a_1}{\beta}|z|^\beta \leq H(z,t) \leq \frac{a_2}{\beta}|z|^\beta \qquad \forall\, z \in \mathbb{R}^{2N}$$

(3) If $z = z(t)$ is a periodic function with minimal period $2\pi q$, $q \in Q$ and $H_z(z(t),t)$ is a periodic function with minimal period $2\pi q$ then necessarily q is an integer.

Remark. The hypothesis (3) is a generic one which implies the essential time dependence of H.

Take for example:

$$H(z,t) = a(t)H(z)$$

where $a(t)$ is a periodic function with minimal period 2π.

Set

$$(1.1) \qquad S_p = \text{least prime factor of } p.$$

We prove:

Theorem 1. Let H satisfy (1), (2) and (3). Then for any integer $p > 1$ satisying:

$$(1.2)_k \qquad \frac{a_2}{a_1} < \left(\frac{2S_p}{k(k+1)}\right)^{\beta/2} \qquad \text{for some integer } k \geq 1;$$

the problem $(1)_p$ admits, at least kN distinct solutions with minimal period $2\pi p$.

If we omit the hypothesis (3) we obtain results of the type of Ekeland-Lasry [2] (see also Girardi-Matzeu [3]); namely:

Theorem 2. Let H satisfy the hypotheses (1) and (2) with $\frac{a_2}{a_1} < 2^{\beta/2}$.

Then for any integer $p > 1$, problem $(1)_p$ admits, at least, N distinct solutions with minimal period $2\pi p$.

In case the Hamiltonian H is independent on the t-variable, i.e. $H = H(z)$, consider the autonomous system:

$$J\dot{z} = H_z(z) \qquad z = (z_1, \ldots, z_{2N})$$

Assume:

(1)' H is convex

(2)' There exists constants a_1, $a_2 > 0$ such that

$$\frac{a_1}{\beta}|z|^\beta \leq H(z) \leq \frac{a_2}{\beta}|z|^\beta \qquad 1 < \beta < 2 \qquad \forall\, z \in \mathbb{R}^{2N}$$

Using theorem 2 one obtains:

Corollary. Let $H = H(z)$ satisfy (1)' and (2)' with $\frac{a_2}{a_1} < 2^{\beta/2}$.

Then for any $T > 0$ there exist, at least, N distict periodic solutions of (1)' with minimal period T.

One may consider the same problem for the second order system of

O.D.E.:

(2) $$\ddot{x} + V_x(x,t) = 0 \qquad x = (x_1,\dots,x_n) \in \mathbb{R}^N$$

where $$V_x(x,t) = (\frac{\partial V}{\partial x_1}(x,t),\dots,\frac{\partial V}{\partial x_n}(x,t)) \quad \text{and}$$

$$V(x,t+2\pi) = V(x,t) \qquad \forall\, x \in \mathbb{R}^n,\ t \in \mathbb{R}.$$

Here the subharmonics are the solutions of the following problem:

$(2)_p$ $$\begin{cases} \ddot{x} + V_x(x,t) = 0 \\ x \quad 2\pi p\text{-periodic} \end{cases} \qquad x = (x_1,\dots,x_n)$$

$p > 1$ integer.

Similarly one obtains:

<u>Theorem 3</u>. Let V satisfy the corresponding assumptions (1), (2) and (3).

Then for any integer $p > 1$ satisfying:

$$\frac{a_2}{a_1} < \left(\frac{3\, S_p^2}{k(k+1)(2k+1)}\right)^{\beta/2}$$

for some integer $k \geq 1$; the problem $(2)_p$ admits, at least, kn distinct solutions with minimal period $2\pi p$.

We shall sketch here the proof of theorem 1.

The proofs of theorem 2 and 3 are obtained in a similar fashion.

Sketch of the proof of theorem 1: it relies on two main facts:

a) The dual variational formulation for the problem $(1)_p$ of Clarke-Ekeland [1]:

solutions of $(1)_p$ are obtained form the critical points of the (dual) functional

$$I^*(u) = \int_0^{2\pi p} \frac{ku\cdot u}{2} - H^*(u,t)$$

defined on the Banach space:

$$E = \{u \in L^{\alpha}[0,2\pi p, \mathbb{R}^{2N}] : \int_0^{2\pi p} u(s)ds = 0\} \qquad \alpha = \frac{\beta}{\beta-1} > 2$$

where

$H^*(\cdot,t)$ = Legendre transform of $H(\cdot,t)$

i.e.

$$H^*(u,t) = \sup_z (u\cdot z - H(z,t))$$

and $K : E \to E$ is the linear (compact) operator that inverts $J\frac{d}{dt}$ on E. That is:

$$Ku(t) = -\int_0^t Ju(s)ds + \frac{1}{2\pi p}\int_0^{2\pi p} dt \int_0^t Ju(s)ds$$

b) The symmetry of the problem $(1)_p$ with respect to a $\mathbb{Z}_p$-group action given by the $2\pi j$-phase shift $j = 0,1,\ldots,p-1$.

Thus to prove existence of multiple critical points for the functional I^* (with minimal period $2\pi p$) one can introduce an appropriate $\mathbb{Z}_p$ index theory.

More precisely define the norm preserving operator:

$$T : E \to E$$

$$Tu(t) = u(t+2\pi) \quad ; \qquad T^p = Id$$

T is the generator of a $\mathbb{Z}_p$ group action on E.

Obviously $I^*(Tu) = I^*(u) \qquad \forall\, u \in E$.

Set

$$\Sigma' = \{A \subset E : A \text{ is closed and } TA = A\}$$

and define the index map:

$$i_p : \Sigma' \to \mathbb{N} \cup \{+\infty\}$$

as follows:

for $A \in \Sigma'$ set $i_p(A) = k$ if k is the smallest non-negative integer such that there exists a continuous map:

$$h: A \to \mathbb{C}^k \setminus \{0\} \qquad h = (h_1, \ldots, h_k)$$

and integers $m_j \neq 0$ relatively prime to the integer p $i \leq j \leq k$, such that:

$$h_j(Tu) = e^{im_j \frac{2\pi}{p}} h_j(u) \qquad j = 1, \ldots, k$$

Set $i_p(A) = 0$ if $A = \emptyset$ and set $i_p(A) = +\infty$ if no such map exists

For $r \in \mathbb{R}$ set

$$A_r = \{ u \in E : I^*(u) \geq r \} .$$

One can show that the index map i_p satisfies the following important property:

for any number α, β with $\alpha < \beta$ such that the interval $[\alpha, \beta]$ does not contain any critical value of I^* we have

$$i_p(A_\alpha) = i_p(A_\beta) .$$

So intuitively if the number

$$r_\ell = \sup \{r : i_p(A_r) = \ell\} = \sup \{ r : i_p(A_r) \geq \ell \} ; \quad \ell \in \mathbb{N}$$

is finite, then it must be a critical value for I^*.

Thus the proof will follow considering that:

(i) the functional I* is bounded from above

(ii) $\forall$ $p > 1$ there exists a constant c_p such that if u is a critical point for I* and $I^*(u) > c_p$ then u has minimal period $2\pi p$.

(iii) If $p > 1$ satisfies $\frac{a_2}{a_1} < \left(\frac{2S_p}{k(k+1)}\right)^{\beta/2}$ then there exists $A_k \in \Sigma$ such that:

$$\inf_{A_k} I^* > c_p \quad ; \quad i_p(A_k) = k\mathbb{N}$$

Finally note that the fix point set F for such $\mathbb{Z}_p$ action is given by

$$F = \{u \in E : u \text{ is } 2\pi\text{-periodic}\}$$

Since we seek critical points for I* with minimal period $2\pi p$ we are sure to avoid the subset F.

This guarantees the claimed multiplicity result.

REFERENCES

[1] Clarke, F., Ekeland, I.: "Hamiltonian trajectories having prescribed minimal period". Comm. Pure Appl. Math. 33 (1980) pp 103 116

[2] Ekeland, I., Lasry, J.M.: "On the number of periodic trajectories for a Hamiltonian flow on a convex energy surface", Annals of Math. 112 (1980) pp 283-319

[3] Girardi, M., Matzeu, M.: "Solutions of minimal period for a class of nonconvex Hamiltonian systems and applications to the fixed energy Problem", Nonlinear Analysis Theory, Methods and Appl. Vol. 10, 4 pp. 371-382 (1986).

BIFURCATION OF INVARIANT TORI FOR NON-HAMILTONIAN SYSTEMS+

Piero Negrini*

The purpose of this note is to illustrate how the methods of the K.A.M. theory can be used in the analysis of the bifurcation phenomena for non hamiltonian systems.

First of all we have to recall that the ideas we shall work with, were introduced by Chencinner in the paper "Bifurcation des points fixes elliptiques-Courbes Invariantes" [1], studying plane maps not measure preserving. Let us now introduce our problem.

We consider the differential system:

$$(1)\qquad \dot{p} = P(\mu,p) = A(\mu)p + \mathbb{P}(\mu,p)$$

where p belongs to a neighbourhood of the origin in $\mathbb{R}^4$, μ is a two dimensional parameter in a neighbourhood of the origin in $\mathbb{R}^2$, P is as smooth as necessary, $\mathbb{P}(\mu,p) = O(|p|^2)$.

We assume that the eigenvalues of A(0) are purely imaginary, i.e. $\sigma(A(0)) = (\pm i\,\omega_1, \pm i\omega_2)$, with $(\omega_1,\omega_2) \in \mathbb{R}^2$, satisfying a "non-resonance condition":

+ Reduced version of the paper [8]

* Present address: D.to Mat. Fis. Università di Camerino, 62032 Camerino (MC) Italy

$$\omega_1 k_1 + \omega_2 k_2 \neq 0 \quad (k_1, k_2) \in Z^2/\{0\}$$

Then, we rewrite (1) in the polar normal form, assuming the transversality hypothesis on the eigenvalues of $A(\mu)$ at $\mu = 0$. Precisely, for any integer ℓ, $\ell \geq 1$, we can find a polynomial transformation of degree 2 +1 (see for instance [2]), which carries system (1) into the system:

$$(1') \quad \begin{cases} \overset{\circ}{r}_i = r_i\{\mu_i + \sum_1^{\ell}{}_k N_{i,(k)}(r_1^2, r_2^2)\} + R_i(r,\theta,\mu) \\ \overset{\circ}{\theta}_i = \omega_i + \sum_1^{\ell}{}_k M_{i,(k)}(r_1^2, r_2^2) + \Theta_i(r,\theta,\mu) \end{cases}$$

where $\mu = (\mu_1, \mu_2)$, $r = (r_1, r_2)$, $r_i \in \mathbb{R}^+$, $\theta = (\theta_1, \theta_2)$, $\theta_i \in S^1$, $i = 1,2$. Moreover $N_{i,(k)}, M_{i,(k)}$ are forms of degree 2k in (r_1, r_2) (with coefficients depending on μ), R_i, $\Theta_i = 0(r^{2(\ell+1)})$. For any ℓ, we consider the "ℓ-reduced system" of (1'):

$$(2) \quad \overset{\circ}{r}_i = r_i\{\mu_i + \sum_1^{\ell}{}_k N_{i,(k)}(r_1^2, r_2^2)$$

Then, under a suitable hypothesis on the coefficients of $N_{i,(\ell)}$ (see [3], [8]) a curve $\Gamma_{(\ell)} = \{(\mu_1, \mu_2),\ \mu_1 = \phi_{(\ell)}(\mu_2)\}$ can be determined, such that the positive fixed point $n_{(\ell)}(\mu)$ (i.e. the zero of the r.h.s. of (2) lying in the domain $r_i > 0, i = 1,2$) becomes a "center in the linear approximation", when considered on $\Gamma_{(\ell)}$. In other words, setting

$$A_{(\ell)}(\mu_2) = \{\partial_{r_j} N_{i(\ell)}(n_{(\ell)}(\mu),\mu)\}_{\mu_1=\phi_{(\ell)}(\mu_2)}, \quad i,j=1,2$$

the "critical curve" $\Gamma_{(\ell)}$ is determined by the conditions:

(3) $\det A_{(\ell)}(\mu_2) > 0$, $\operatorname{Tr} A_{(\ell)}(\mu_2) = 0$

One can easily verify (see [3]) that for $\ell = 1$ $n_{(1)}(\mu_2) =: n_{(1)}(\mu_1 = \phi_{(1)}(\mu_2), \mu_2)$, is a center for the corresponding system.

Assume now that $\ell = 2$, $\mu_1 = \phi_{(2)}(\mu_2)$. The behaviour of the flow close to $n_{(2)}(\mu_2)$ depends on the non linear terms (with respect to the variable $x = r - n_{(2)}(\mu_2)$). By means of an algebraic procedure (the Poincaré-Liapunov procedure, (see [4]), we can give sufficient conditions in order that $n_{(2)}(\mu_2)$ is a 3-attractor (or a 3-repellor).

Setting now $\mu_1 = \phi_{(\ell)}(\mu_2) + \nu$, we cross the critical curve $\Gamma_{(\ell)}$, and consider the positive fixed point $n_{(\ell)}(\mu_2, \nu)(n_{(\ell)}(\mu_2, 0) = n_{(\ell)}(\mu_2))$. By choosing the sign of ν, we can obtain that $n_{(\ell)}(\mu_2, \nu)$ is an hyperbolic repellor when $n_{(2)}(\mu_2)$ is a 3-attractor (or viceversa). Therefore the usual Hopf bifurcation occurs, and the bifurcating cycles $C_{\mu_2, \nu}$ are hyperbolic attractors (resp. hyperbolic repellors). Obviously, for the system obtained from (1') forgetting the $O(r^{2(\ell+1)})$ terms, the family of fixed points $n_{(\ell)}(\mu_2, \nu)$ of (2) represents a family of 2-dimensional invariant tori; the bifurcating cycles $C_{\mu_2, \nu}$ correspond to 3-dimensional invariant tori.

Now let be $\ell \geq 12$. By a scaling of variables and some non linear transformations (sending $n_{(\ell)}(\mu_2, \nu)$ into 0) we rewrite (1') as

$$\dot{y} = [\beta(\mu_2, \delta)J + \alpha(\mu_2, \delta)]y + \mu_2^{\ell/2}\{|y|^2(a(\mu_2) + b(\mu_2)J)y + \mu_2^{\ell/2}Y((y, \theta\mu_2, \delta)\}$$

(4)

$$\dot{\theta} = \omega(\mu_2, \delta) + \mu_2^{1+\ell/2}\{\gamma(\mu_2)|y|^2 + \mu_2^{\ell/2}\tilde{\Theta}(y, \theta\mu_2, \delta)\}$$

(where, without loss of generality, we assume $\mu_2 > 0$). J is the 2×2 symplectic matrix, $\nu = \mu_2^{(\ell+4)/2}$; the constants $\beta(\mu_2, \delta)$, $\alpha(\mu_2, \omega)$, $\omega(\mu_2, \delta)$ are determined from the linear analysis of system (2) w.r. to the fixed point $n_{(\ell)}(\mu_2, \nu)$, and the constants $a(\mu_2)$, $b(\mu_2)$, $\gamma(\mu_2)$ are identified analysing the non linear terms (w.r. to $x=r-n_{(\ell)}(\mu_2))$ of system (2) at $\mu_1 = \phi_{(\ell)}(\mu_2)$. In particular, it turns out that $a(\mu_2) = \mu_2^2 g(\mu_2)$, $g(0)$ being computed with the foresaid Poincaré-Liapunov procedure.

Now let δ be such that

$$(5) \qquad 0 < -[\partial_\delta \alpha(0,0)]\delta / g(0) (:= s_o^2).$$

Then we set $y = s(\cos\phi, \sin\phi)$, $s = s_o + \mu_2^3\sigma$, so that (4) becomes:

$$(6) \qquad \begin{aligned} \dot{\sigma} &= \mu_2^{(\ +4)/2}\{[-[\partial_\delta \alpha(0,0)]\sigma]\sigma + O(\mu_2) \\ \dot{\phi} &= \beta(\mu_2,\delta) + O(\mu_2^{\ /2+3}) \\ \dot{\theta} &= \omega(\mu_2,\delta) + O(\mu_2^{\ /2+3}) \end{aligned}$$

where the unspecified $O(\mu_2^{\ell/2+3})$ terms depend on $(\sigma,\phi,\theta,\mu_2,\delta)$. Now, using the results in [5], we can prove the following theorem:

<u>Theorem</u>. There exist two positive real numbers μ_o,δ_o, such that if $0 < \mu_2 < \mu_o$, $\delta_o/2 < |\delta| < \delta_o$, and δ satisfies (5), then system (6) admits a family of 3-invariant tori $\Pi^3_{\mu_2,\delta}$ close to $\sigma = 0$ (all hyperbolic attractors (resp. repellors) if $g(0) < 0$ (resp. if $g(0) > 0$). (For a deeper result on the existence of 3-quasi periodic solutions see [6]).

Now, in order to prove that $\Pi^3_{\mu_2,\delta}$ bifurcate from 2-invariant tori, we have to study the system (4) as $\delta \to 0$, therefore we leave the "domain of the hyperbolic theory". The mean role will be played now by the rotations, as well as in the case of Hamiltonian systems. We approach this bifurcation problem, by using a procedure of the K.A.M.. type, based on a paper of Braaksma and Broer ([7]). More precisely we improve their results in order to discuss problems of the form:

$$\dot{x} = \lambda x + X_o(x,\theta)$$

(7)

$$\dot{\theta} = \omega + \Theta_o(x,\theta)$$

where $x \in \mathbb{C}$, $\theta \in \Pi^2$, $\lambda = \lambda_1 + i\lambda_2$ (in [7]) $\Theta_o(x,\theta) = 0$). We do this, having in mind the system obtained from (4) setting $x = y_1 + iy_2$, $\lambda = \alpha(\mu_2,\delta) + i\beta(\mu_2,\delta)$.

We work under analyticity assumptions on the r.h.s. of (7) on a domain $0 = 0_o \times \Lambda_1 \times \Lambda_2 \times \Omega$, $x \in 0_o$, $\lambda_1 \in \Lambda_1$, $\lambda_2 \in \Lambda_2$, $\omega \in \Omega$, Λ_1 including the zero, Λ_2, Ω are bounded away from zero. Given a positive γ -sufficiently small- and a real $\tau, \tau \geq 2$, we introduce "the set of frequencies" $C_\gamma = \{(\omega,\lambda_2) \in \Omega \times \Lambda_2 : |(\omega,j) - \ell\lambda_2| \geq 2\gamma/|j|^\tau,\ j \in Z^2/\{0\}, \ell \in Z, |\ | \leq 4\}$. The purpose is to continuate the invariant torus $x = 0$ of system obtained from (7) setting $X_o(x,\theta) = \Theta_o(x,\theta) = 0$, to the full system, when (λ_2,ω) are "suitable deformations" of $(\lambda_2^*,\omega^*) \in C_\gamma$ (provided that $X_o\Theta_o$ are sufficiently small).

Let us state the principal steps of the procedure. For any integer i, $i \geq 1$, we construct a suitable complex domain $W^{(i)}$ (such that $W^{(i)}$ "collapses" to $\{0\} \times \Pi^2 \times \Lambda_1 \times C_\gamma$ as $i \to \infty$), and a map Ψ_{i-1} ($\Psi_{i-1}(W^{(i)}) \subset W^{(i-1)}$)

$$\Psi_{i-1}: (\zeta_i, \phi_i, \lambda_i, \omega_i) \to (\zeta_{i-1}, \phi_{i-1}, \lambda_{i-1}, \omega_{i-1})$$

(where $\zeta_o = z$, $\phi_o = \theta$, $\lambda_o = \lambda$, $\omega_o = \omega$), of the form:

$$\begin{cases} \zeta_{i-1} = \zeta_i + \sum\limits_{0 \leq k_1+k_2 \leq 3} u_i(\phi_i, \lambda_i, \omega_i)\zeta_i^{k_1}\bar{\zeta}_i^{k_2} \\ \phi_{i-1} = \phi_i + \sum\limits_{0 \leq k_1+k_2 \leq 3} v_i(\phi_i, \lambda_i, \omega_i)\zeta_i^{k_1}\bar{\zeta}_i^{k_2} \\ \lambda_{i-1} = \lambda_i + U_i(\lambda_i,\omega_i) \\ \omega_{i-1} = \omega_i + V_i(\lambda_i,\omega_i) \end{cases}$$

which sends the system:

$$\dot{\zeta}_{i-1} = \lambda_{i-1}\ \zeta_{i-1} + a_{i-1}(\lambda_{i-1}, \omega_{i-1})\zeta_{i-1}|\zeta_{i-1}|^2 + X_{i-1}(\zeta_{i-1}, \phi_{i-1})$$

$$\dot{\phi}_{i-1} = \omega_{i-1} + b_{i-1}(\lambda_{i-1}, \omega_{i-1})|\zeta_{i-1}|^2 + \Phi_{i-1}(\zeta_{i-1}, \phi_{i-1})$$

in the corresponding system where i-1 is replaced by i and

$$\max\{\sup|X_i|_{W^{(i)}}, \sup|\Phi_i|_{W^{(i)}}\} \leq \varepsilon_i \max\{\sup|X_{i-1}|_{W^{(i)}}, \sup|\Phi_{i-1}|_{W^{(i)}}\}$$

ε_i decreasing very fast as $i \to \infty$ ($a_\circ = b_\circ = 0$, $\Phi_\circ = \Theta_\circ$).

The functions u_i, v_i are quasi periodic and are determined together with U_i, V_i, a_i, b_i, by solving linear partial differential equations (the "homological equations") (see [8]).

Then, by using the "limit" map $\phi = \lim_{i\to\infty} \phi_i$, $\phi_i := \Psi_\circ \circ \Psi_1 \ldots \Psi_{i-1}$: $W^{(i)} \to W^{(0)}$ we can construct a transformation which is of the form:

$$(8)\qquad \begin{cases} z = \zeta + \sum\limits_{0\leq k_1+k_2\leq 3} L_{k_1,k_2}(\phi, \lambda_*, \omega_*)\zeta^{k_1}\bar{\zeta}^{k_2} \\ \theta = \phi + \sum\limits_{0\leq k_1+k_2\leq 3} M_{k_1,k_2}(\phi, \lambda_*, \omega_*)\zeta^{k_1}\bar{\zeta}^{k_2} \\ \lambda = \lambda_* + U(\lambda_*, \omega_*) \\ \omega = \omega_* + V(\lambda_*, \omega_*) \end{cases}$$

((λ_{*2}, ω_*) $\mathcal{C}_\gamma$, U,V being as small as $X_\circ, \Theta_\circ$), by means of which we conjugate system (7) to the system:

$$\begin{cases} \dot{\zeta} = \lambda_*\zeta + a_*(\lambda_*, \omega_*)\zeta|\zeta|^2 + O(|\zeta|^4) \\ \dot{\phi} = \omega_* + b_*(\lambda_*, \omega_*)\ |\zeta|^2 + O(|\zeta|^4). \end{cases}$$

Therefore the invariant 2-dimensional tori of (7) are obtained from (8), setting $\zeta = 0$; moreover the asymptotic properties of $\zeta = 0$ at $\mathbb{R}e\lambda_* = 0$ are given by the sign of $\alpha_* := \{\mathbb{R}e\, a_*(\lambda_*, \omega_*)\}_{\mathbb{R}e\lambda_*=0}$.

We came back now to the system (4), observing that the control on the corresponding X_o, Θ_o, is obtained by taking ℓ sufficiently large. Moreover it is possible to prove that α_* is "well estimated" by $\mu_2^2 g(0)$. Finally, taking into account that $\beta(\mu_2, 0) = \mu_2\{\beta_o + O(\mu_2)\}$, $\omega(\mu_2, 0) = \omega + O(\mu_2)$, we can determine from the last two rows of (8) the points $(\mu_2^{(c)}, \mu_2^{(c)})$ for the crossing of the critical curve.

In order to identify the "good bifurcation paths" we use a consequence of the esistence of ϕ. More precisely, one uses the fact that system (7) can be conjugated to the system:

$$\begin{cases} \dot{\zeta} = \chi\zeta + a_*(\chi,\tilde{\omega})\zeta|\zeta|^2 + R(\zeta,\phi,\chi,\tilde{\omega}) + Q(\zeta,\phi,\chi,\tilde{\omega}) \\ \\ \dot{\phi} = \tilde{\omega} + b_*(\chi,\tilde{\omega})\ |\zeta|^2 + R_1(\zeta,\phi,\chi,\tilde{\omega}) + Q_1(\zeta,\phi,\chi,\tilde{\omega}) \end{cases}$$

where $\chi = \chi_1 + i\chi_2$, $R, R_1 = O(|\zeta|^3)$, Q, Q_1 are identically zero on $\Lambda_1 \times C_\gamma$.

Then for any $(\lambda_{*2}, \omega_*) \in C_\gamma$ we define the cuspidal domain:

$$C_{\lambda_{*2},\omega_*} := \{(\chi_1,\chi_2,\tilde{\omega}) \in \Lambda_1\times\Lambda_2\times\Omega :\ |\chi_2-\lambda_{*2}|^n, |\tilde{\omega}-\omega_*|^n \leq \frac{1}{L}|\chi_1|\}$$

with L a real positive number (sufficiently large), n an arbitrary positive integer. Inside this domain we are able to use again the hyperbolic theory to show that the bifurcation of 2-dimensional invariant tori into 3-dimensional invariant tori occurs along any path crossing the vertex $\chi_1 = 0$.

REFERENCES

[1] Chencinner, A.: "Bifurcation des points fixes elliptiques. I. Courbes Invariantes", I.H.E.S., n. 61, (1986)

[2] Chow, S.N., Hale, J.K.: "Methods of Bifurcation Theory", Springer Verlag (1982)

[3] Guckenheimer, J., Holmes, P.: "Nonlinear Oscillations: Dynamical Systems and Bifurcation of Vector Fields", Springer-Verlag (1983)

[4] Negrini, P., Salvadori, L.: "Attractivity and Hopf Bifurcation", Non Linear Analysis, T.M.A. 3, I (1979)

[5] Sacker, R.J.: "A new approach to the perturbation theory of invariant surfaces", Comm. Pure Appl. Math. 18 (1965)

[6] Braaksma, B.L.J., Broer, H.W.: "Quasi periodic flow near a condimension one singularity of a divergence free vector field in dimension 4", Asterisque, n. 98-99 (1981)

[7] Braaksma, B.L.J., Broer, H.W.: "On a quasi-periodic Hopf bifurcation", Preprint

[8] Maffei, C., Negrini, P., Scalia, M.: "Bifurcation from 2-dimensional to 3-dimensional invariant tori", Preprint.

ON THE STRUCTURE OF THE LARGEST BIFURCATING SETS
OF PERIODIC DIFFERENTIAL SYSTEMS

L. Salvadori
Dipartimento di Matematica, Università di Trento
Povo (Trento), Italy

1. Introduction

We are concerned with bifurcation and related stability problems of one parameter family of differential systems

(1.1)$_\mu$ $\quad \dot{u} = f(t,u,\mu)$,

where $f \in C^1(\mathbb{R} \times \mathbb{R}^n \times \mathbb{R}, \mathbb{R}^n)$, $f(T,0,\mu) \equiv 0$, and $f(t+T,u,\mu) = f(t,u,\mu)$ for some fixed $T > 0$. In [5] it was provided information on the existence and stability properties of invariant periodic sets $M\mu$ in $\mathbb{R} \times \mathbb{R}^n$ bifurcating from the equilibrium configuration $u = 0$. The occurrence of bifurcation is only due to a change of stability of equilibrium, namely from asymptotic stability to instability, when μ crosses a critical value ($\mu = 0$). Each section $M\mu(t_o)$ of $M\mu$ with the plane $t = t_o$ is the largest compact subset of $\mathbb{R}^n$ disjoint from the origin (contained in a fixed neighbourhood of the origin) which is invariant under the autonomous discrete dynamical system $\pi_\mu^{t_o}$ that naturally arises from (1.1)$_\mu$.

The structure of the sets M_μ is determined by further assumptions which complement the above stability requirements. Assume that $m \leq n$ Floquet multipliers of $D_u f(t,0,\mu)$ are on the unit circle c for $\mu = 0$

and outside of c for $\mu > 0$, and the remaining n-m multipliers are inside c. Then by an appropriate transformation of coordinates which involves periodically t, the center manifold theorem reduces the inspection of $M\mu$ to the analysis of an m-dimensional differential system, $R_m(\mu)$, whose linear parts are t-independent. The case m = 1 was treated in [5], and the results are briefly described here.

In the present work we are more concerned with the case m = 2. Extensive proofs will appear in a forthcoming paper with S. Bernfeld and F. Visentin. The eigenvalues, $\alpha(\mu) \pm i\beta(\mu)$, of $R_2(\mu)$ satisfy $\alpha(0) = 0$ and $\alpha(\mu) > 0$ for $\mu > 0$; in addition we assume $\beta(0) \neq 0$. We suppose that $R_2(0)$ is autonomous and that the asymptotica stability of the origin of $R_2(0)$ is preserved when we perturb $R_2(0)$ by terms of order greater than some integer h. Different situations arise. They depend upon: (1) the value of h (which in any case has to be odd and ≥ 3); (2) whether a resonance condition between T and $2\pi/\beta(0)$ is satisfied or not; (3) whether $\alpha'(0) > 0$ (transversality condition) or $\alpha'(0) = 0$. The results are obtained by modifying and extending to the above discrete dynamical systems $\pi_\mu^{t_o}$ a procedure done by Lanford [4] for Poincaré maps in problems dealing with the secondary Hopf bifurcation of autonomous differential systems.

In the last part of this paper we explore the existence and stability of periodic solutions lying on $M\mu$. If these solutions exist, a resonance condition must be satisfied and under such a condition we provide a complete characterization of existence in the generic case (h = 3). The stability properties of the periodic solutions are analyzed by investigating the stability behaviour of the fixed points of $\pi_\mu^{t_o}$ lying on $M\mu(t_o)$. In particular we discuss the possible configurations of attractors, repellors and semistable fixed points and provide methods to distinguish the different possibilities.

2. Exchange of stability and largest bifurcating sets

Let M be a set in $R \times \mathbb{R}^n$ such that for every $t \in \mathbb{R}$ the set $M(t) = \{u \in \mathbb{R}^n : (t,u) \in M\}$ is not empty. If there exists a compact set Q in $\mathbb{R}^n$ satisfying $M(t) \subset Q$ for all $t \in \mathbb{R}$, then M is said to be s-bounded. If in addition each M(t) is compact, then we say that M is s-compact. When the mapping $t \to M(t)$ is periodic or in particular t-independent, we will say that M is periodic or t-independent respectively.

Consider now the differential systems $(1.1)_\mu$. We are only interested to the behaviour of the flow near the origin in $\mathbb{R}^n$. Therefore, we may assume without any loss of generality (by a suitable redefinition of f outside of a neighbourhood of the origin), that for each $\mu, t_o \in \mathbb{R}$ and $u_o \in \mathbb{R}^n$, the solution $u(t,t_o,u_o,\mu)$ through (t_o,u_o) exists for all $t \in \mathbb{R}$. We call $(1.1)_o$ the unperturbed system. We will denote by M_o the set in $\mathbb{R} \times \mathbb{R}^n$ $\{(t,u): t \in \mathbb{R},\ u=0\}$. For any fixed t_o, μ consider the map $\pi = \pi_\mu^{t_o}: \mathbb{Z} \times \mathbb{R}^n \to \mathbb{R}^n$ defined by $\pi(i,u) = u(t_o + i\,T, t_o, u, \mu)$. Clearly we have $\pi(0,u) = u$ and $\pi(i_1+i_2,u) = \pi(i_2, \pi(i_1,u))$ for every $i_1, i_2 \in \mathbb{Z}$ and $u \in \mathbb{R}^n$. Hence π defines an autonomous discrete dynamical system depending on t_o, .

<u>Definition 2.1</u>. We say that $\mu = 0$ is a bifurcation value on the right for the family $(1.1)_\mu$ if there exists $\bar{\mu} > 0$ and a family $(M\mu)$, $\mu \in (0, \bar{\mu})$, of s-compact subsets of $(\mathbb{R} \times \mathbb{R}^n) \setminus M_o$ having the following properties:

(a) for each $\mu \in (0, \bar{\mu})$ $M\mu$ is a T-periodic invariant set of $(1.1)_\mu$ and $M_\mu(t) \neq \Phi$ for all $t \in \mathbb{R}$;

(b) $M\mu(t) \to \{0\}$ as $\mu \to 0$ uniformly in t.

<u>Theorem 2.1</u>. Suppose that $u(t) \equiv 0$ is an asymptotically stable solution of $(1.1)_o$ and a completely unstable (i.e. asymptotically stable in the past) solution of $(1.1)_\mu$ for $\mu > 0$ small. Then $\mu = 0$ is a

bifurcating value on the right. Precisely there exist $\bar{\mu} > 0$ and an s-compact neighbourhood H of M_o such that for each $\mu \in (0,\bar{\mu})$ the largest s-compact invariant set of $(1.1)_\mu$ contained in $H \setminus M_o$, say M_μ, is non-empty, T-periodic, and the family (M_μ) satisfies (b) in Definition 2.1. Moreover each M_μ is an asymptotically stable set of $(1.1)_\mu$.

The proof of Theorem 2.1 was given in [5] by using an appropriate periodic Liapunov function associated with the asymptotic stability of the null solution of $(1.1)_o$ as well as the properties of the dynamical systems $\pi_\mu^{t_o}$. Assume now that $m \leq n$ Floquet multipliers of $D_u f(t,0,\mu)$ are on the unit circle c for $\mu = 0$ and outside of c for $\mu > 0$ and that the remaining n-m multipliers are inside c. Then by a suitable linear transformation of coordinates (whose coefficients depend on μ and periodically on t), $(1.1)_\mu$ may be written as

$$\dot{p} = A(\mu)p + P(t,p,q,\mu)$$
$$\dot{q} = B(\mu)q + Q(t,p,q,\mu)$$

where $p \in \mathbb{R}^m$, $q \in \mathbb{R}^{n-m}$ and $A(\mu)$, $B(\mu)$ are $m \times m$ and $(n-m) \times (n-m)$ t-independent matrices respectively. The eigenvalues of $A(\mu)$ have real parts for $\mu = 0$ and positive real parts for $\mu > 0$; the eigenvalues of $B(\mu)$ have negative real parts. Moreover P and Q have at $p = 0$, $q = 0$ a zero of order greater than 1. Due to the local character in u,μ of our problems, given any $\sigma > 0$ we may assume without loss of generality that the norm of $W = (P,Q)$ in the C^1 topology satisfies $|W| < \sigma$. Then if $\mu^* > 0$ and σ are sufficiently small, there exists a function ϕ : $\mathbb{R} \times \mathbb{R}^m \times [0,\mu^*] \to \mathbb{R}^{n-m}$, $\phi \in C^1, \phi(t,0,\mu) \equiv 0$ which is T-periodic in t such that for every $\mu \in [0,\mu^*]$ the set

(2.1) $\qquad S_\mu = \{(t,p,q) : q = \phi(t,p,\mu)\}$

is an invariant manifold of $(1.1)_\mu$, and

$$\| q(t,t_o,p_o,q_o,\mu) - \phi(t,p(t,t_o,p_o,q_o),\mu) \| \leq L\, e^{-\beta(t-t_o)} \quad t \geq t_o$$

where $\beta > 0$, $L > 0$ are constant and L depends continuously on t_o, p_o, q_o ([3]; see also [2]). Hence we have the following result.

Theorem 2.2. Suppose that $p(t) \equiv 0$, $q(t) \equiv 0$ is an asymptotically stable solution of $(1.1)_o$. Then the conclusion of Theorem 2.1 holds and each bifurcating set M_μ lies on the invariant manifold S_μ.

3. Largest bifurcating sets in cases m = 1 and m = 2

We recall a concept of asymptotic stability that we have used elsewhere for autonomous as well as for periodic differential systems. Let Γ_τ^k, $\tau \geq 0$, $k \geq 1$, be the set of functions $W : \mathbb{R} \times \mathbb{R}^n \to \mathbb{R}^n$ such that: (1) W(t,u) has continuous partial derivatives in u up through order k; (2) W is periodic in t (of period independent of u) and satisfies $\tau = \inf\{\lambda > 0: W \text{ is } \lambda\text{-periodic in } t\}$. Consider the n-dimensional system $\dot{u} = W(t,u)$ where $W \in \Gamma_\tau^k$ and $W(t,0) \equiv 0$. Let h, $0 < h \leq r$, be an integer. The solution $u(t) \equiv 0$ is said to be h-asymptotically stable if: (i) for any $\zeta \in \Gamma_\tau^k$ and $\zeta = o(\| u \|^h)$ as $u \to 0$, the solution $u(t) \equiv 0$ of the system $\dot{u} = W(t,u) + \zeta(t,u)$ is asymptotically stable; (ii) property (i) is not satisfied when h is replaced by any $\nu = 1,\ldots$ h-1.

The aim of this section is to analyze the bifurcating periodic set M_μ of Theorem 2.2 in the two cases that m = 1 and m = 2.

(a) m = 1. System $(1.1)_\mu$ becomes

$$(3.1)_\mu \qquad \begin{aligned} \dot{p} &= \alpha(\mu)p + P(t,p,q,\mu) \\ \dot{q} &= B(\mu)q + Q(t,p,q,\mu) \end{aligned}$$

where $p \in \mathbb{R}$, $q \in \mathbb{R}^{n-1}$, $\alpha(\mu) \in \mathbb{R}$, and $B(\mu)$ is a $(n-1)\times(n-1)$ matrix whose eigenvalue have negative real parts. Moreover $\alpha(0) = 0$ and $\alpha(\mu) > 0$ when $\mu > 0$. The following theorem holds.

Theorem 3.1. Suppose that $p(t) \equiv 0$, $q(t) \equiv 0$ is an asymptotically stable solution of $(3.1)_o$. Then each bifurcating set M_μ of Theorem 2.2 is the union of two nonempty disjoint connected s-compact sets, M_μ^-, M_μ^+, which are invariant and asymptotically stable under $(3.1)_\mu$. For each t the sections $M_\mu^-(t)$, $M_\mu^+(t)$ are arcs of $S_\mu(t)$ (which is now unidimensional) and are located in the regions $p < 0$ and $p > 0$ respectively. The endpoints (which may coincide) are fixed points of π_μ^t and correspond to T-periodic trajectories of $(3.1)_\mu$. Therefore if in M^- (resp. M^+) there exists only one bifurcating T-periodic trajectory, then M_μ^- (resp. M_μ^+) reduces to this trajectory which will be asymptotically stable.

An example of the last situation considered in Theorem 3.1 may be obtained by strengthening the assumption of asymptotic stability of the null solution of $(3.1)_o$. Precisely we have:

Theorem 3.2. Suppose that $P,Q \in C^k$, $k \geq 3$, and that one of the following hypotheses is satisfied: (1) the origin of $(3.1)_o$ is 3-asymptotically stable; (2) the origin of $(3.1)_o$ is h-asymptotically stable with $3 < h \leq k$, and $\alpha'(0) \neq 0$. Then each one of the two sets M_μ^-, M_μ^+ reduces to a T-periodic asymptotically stable trajectory of $(3.1)_\mu$.

(b) m = 2. By a convenient change of coordinates, system $(1.1)_\mu$ restricted to S_μ may be written as

$$(3.2)_\mu \qquad \begin{aligned} \dot{x} &= \alpha(\mu)x - \beta(\mu)y + X(t,x,y,\mu) \\ \dot{y} &= \alpha(\mu)y + \beta(\mu)x + Y(t,x,y,\mu) \end{aligned}$$

where $\alpha(0) = 0$, $\alpha(\mu) > 0$ for $\mu > 0$, and X,Y have at $x = y = 0$ a zero of order greater than 1. Moreover we assume $\beta(0) \neq 0$.

Theorem 3.3. Suppose that $X,Y \in C^k$ for some $k \geq 3$, that $(3.2)_\circ$ is autonomous and the origin of $(3.2)_\circ$ is h-asymptotically stable, $3 \leq h \leq k$. Assume the transversality condition $\alpha'(0) > 0$. Then each section $M_\mu(t)$ of the bifurcating set M_μ of Theorem 2.2 is a Jordan curve encircling the origin represented in polar coordinates by $r = g_\mu(\theta,t)$ where $g_\mu \in C^k$ is 2π-periodic in θ and T-periodic in t.

Thus the bifurcating sets M_μ are two-dimensional tori in $\mathbb{R} \times \mathbb{R}^n$ (by interpreting t as an angular variable). The same conclusion may be obtained at least in case $h = 3$ (generic case) when the above transversality condition does not hold provided $\alpha(\mu)$ is not a flat function and an appropriate non resonance condition is satisfied. Precisely we have:

Theorem 3.4. Suppose that $X,Y \in C^3$, that $(3.2)_\circ$ is autonomous and the origin of $(3.2)_\circ$ is 3-asymptotically stable. Assume that $\alpha(\mu) = a\mu^s + o(\mu^s)$, $a > 0$, $s > 1$, and $e^{i\nu\beta(0)T} \neq 1$, $\nu = 1,3$. Then the conclusion of Theorem 3.3 holds.

We give an outline of the proof of Theorem 3.3. To this end some preliminaries are needed. Under the assumptions of Theorem 3.3, by using polar coordinates system $(3.2)_\mu$ may be written as

$$(3.3)_\mu \qquad \begin{aligned} \dot{r} &= a\mu r - gr^h + r^{h+1}\gamma(\theta,r) + \mu r^2\delta(t,\theta,t,\mu) + \mu^2 r\chi(\mu) \\ \dot{\theta} &= \beta(\mu) + r^{h-1}\phi(\theta,r) + \mu r\psi(t,\theta,r,\mu) \end{aligned}$$

where $a > 0$, $g > 0$ are constants and $\gamma,\delta,\chi,\phi,\psi$ are C^k functions.

Lemma 3.5. There exist $\bar{r} > 0$, $\bar{\mu} > 0$ such that for every $\mu \in (0,\bar{\mu})$

it is possible to determine $r_1, r_2 \in (0,\bar{r})$, $r_1 < r_2$, such that the annulus $A_\mu = \{(\theta,r): r_1 \leq r \leq r_2\}$ is an asymptotically stable set of $(3.2)_\mu$ and its region of attraction contains the set $\{(r,\theta): 0 < r < \bar{r}\}$. Precisely we can assume $r_1 = b(\mu^p - \mu^q)$, $r_2 = b(\mu^p + \mu^q)$, with $p = (h-1)^{-1}$, $b = (a/g)^p$, $q = p + c$, $c \in (0,p)$.

<u>Outline of the proof of Theorem 3.3</u>. For given t_o,μ let $\pi_\mu^{t_o}$ be the discrete dynamical system generated by $(3.2)_\mu$. By the change of coordinates $z = (r - b\mu^p)/b\mu^q$ (b,p,q are defined as in Lemma 3.5), system $(3.3)_\mu$ becomes

$$\text{(3.4)} \qquad \begin{aligned} \dot{z} &= a(1-h)\,z + \mu^\lambda \Phi(t,\theta,z,\mu) \\ \dot{\theta} &= \beta_1(\mu) + \mu^\lambda \Psi(t,\theta,z,\mu) \end{aligned}$$

where $\lambda \in (1,2)$, and the annulus A_μ of Lemma 3.5 is represented by the set $\{(\theta,z): |z| \leq 1\}$. By integrating over $[t_o, t_o + T]$ we determine $\pi_\mu^{t_o}(1,\cdot)$ in terms of coordinates θ, z. Setting $(\theta,z) = \pi_\mu^{t_o}(1,(\theta_o,z_o))$, we have

$$\begin{aligned} z &= [1 - (h-1)a\,T\mu]\,z_o + \mu^\lambda H(t_o,\theta_o,z_o,\mu) \\ \theta &= \theta_o + \beta_1(\mu)T + \mu^\lambda K(t_o,\theta_o,z_o,\mu) \end{aligned}$$

Denote by U the set of all functions $u \in C^k(\mathbb{R},\mathbb{R})$ which satisfy: (i) $u(\theta + 2\pi) = u(\theta)$ and $|u(\theta)| \leq 1$ for all θ; (ii) $|u(\theta)-u(\theta')| \leq |\theta - \theta'|$ for all θ,θ'. By the same arguments used by Lanford in [4], it is easy to prove that for every θ_o there is a unique $\tilde{\theta}_o$ such that

$$\theta_o = \tilde{\theta}_o + \beta_1(\mu)T + \mu^\lambda K(t_o,\tilde{\theta}_o,u(\tilde{\theta}_o),\mu) \qquad (\text{mod } 2\pi),$$

and that the map $\mathcal{F}: U \to U$ defined by

$$(\mathscr{F}u)(\theta_\circ) = [1 - (h - 1)a\, T\, \mu]\, u(\tilde{\theta}_\circ) + \mu^\lambda H(t_\circ,\tilde{\theta}_\circ,u(\tilde{\theta}_\circ),\mu)$$

is a contraction. Then the manifold $\Gamma_\mu(t_\circ)$ corresponding to the unique fixed point of $\mathscr{F}$ is an invariant attractor under $\pi_\mu^{t_\circ}$, and the region of attraction contains A_μ. Hence the properties of A_μ and $M_\mu(t_\circ)$ imply $\Gamma_\mu(t_\circ) = M_\mu(t_\circ)$. The equation of $M_\mu(t_\circ)$ in polar coordinates will be then $r = g_\mu(\theta_\circ,t_\circ)$ where $g(\theta_\circ,\cdot)$ is C^k and 2π-periodic in $\theta_\circ$. The proof that g_μ is T-periodic in $t_\circ$ and $g_\mu \in C^k$ (in the pair $\theta_\circ,t_\circ$) follows from the properties of M_μ. Indeed the mapping $t \to M_\mu(t)$ is T-periodic and $M_\mu(t)$ is the image of $M_\mu(0)$ under the flow $(3.2)_\mu$.

If $h = 3$, if $\alpha(\mu)$ satisfies the condition in Theorem 3.4 for some $s > 1$, and if in $(3.2)_\mu$ the terms of order 2 in x,y are absent for all μ, then the conclusion of Theorem 3.3 holds. Indeed slight modifications of Lemma 3.5 and corresponding changes in the definition of z allows system $(3.2)_\mu$ to assume the form $(3.4)_\mu$ except that μ is replaced by μ^s in the linear part, and $\lambda > s$. The remainder of the proof is similar to that of Theorem 3.3. The non resonance condition in Theorem 3.4 is just sufficient for the existence of a suitable transformation of coordinates x,y which eliminates the above terms of order 2 without modifying the linear part as well as the essential requisites of $(3.2)_\mu$.

Finally we observe that Theorems 3.3 and 3.4 hold even if $(3.2)_\circ$ is not restricted to be autonomous provided an appropriate non resonance condition is satisfied. When $h = 3$ this condition is $e^{i\nu\beta(0)T} \neq 1$ $\nu = 1,2,3,4$.

4. Periodic solutions of planar systems

We are interested in determining the existence and stability of nontrivial periodic solutions of $(3.2)_\mu$ in a neighbourhood of the origin. Clearly the trajectories of these solutions lie on M_μ and for any given t_o they correspond to the fixed points of $\pi_\mu^{t_o}$ on $M\mu(t_o)$. When the nontrivial periodic solutions exist then necessarily $\beta(0)$ must satisfy the resonance condition $\beta(0) = 2\pi\nu/T$ for some integer ν (we assume $\nu = 1$). Indeed if $\beta(0) \neq 2\pi\nu/T$ then an application of the implicit function theorem implies the existence of a unique periodic solution near the origin, which must be the zero solution. In the study of the secondary bifurcation from a periodic solution into tori a non resonance condition is required (see [2] and references within), thus eliminating the possibility of having periodic solutions lying on the tori.

Throughout this section we will assume that: (a) $\alpha'(0) > 0$; (b) $\beta(0) = 2\pi/T$; (c) $X, Y \in C^k$ for some $k \geq 3$; (d) the unperturbed system $(3.2)_o$ is autonomous and the origin of $(3.2)_o$ is h-asymptotically stable, $3 \leq h \leq k$. Consider the system associated with (3.2)

(S)
$$\dot{x} = \alpha(\mu)x - \varepsilon y + X(t,x,y,\mu)$$
$$\dot{y} = \alpha(\mu)y + \varepsilon x + Y(t,x,y,\mu)$$

where ε will be treated as a second parameter. Clearly if $h = 3$, the zero solution of (S) when $\mu = 0$ is 3-asymptotically stable for all ε. Then, by a general result in [ ] we have in case $h = 3$ that there exist $\bar{c}, \bar{\mu}, \bar{\varepsilon} > 0$ and two C^1 functions $c(s,\mu)$, $\varepsilon(s,\mu)$ which are defined in $\mathbb{R} \times [0,\bar{\mu}]$ and satisfy: (1) $c(s,0) = 0$, $c(s,\mu) > 0$ for $\mu > 0$, and $\varepsilon(s,0) = 2\pi/T$; (2) if $s \in \mathbb{R}$, $c \in [0,\bar{c}]$, $|\varepsilon - 2\pi/T| \leq \bar{\varepsilon}$, and $\mu \in [0,\bar{\mu}]$, then the solution of (S) through $t_o = s$, $x_o = c$, $y_o = 0$ is T-periodic if and only if $c = c(s,\mu)$, $\varepsilon = \varepsilon(s,\mu)$. Thus we have the following characterization of the existence and the number of nontrivial periodic solution of $(3.2)_\mu$.

Theorem 4.2. Assume $h = 3$. Let $A\mu$ be the set of all nonzero T-periodic solutions of $(3.2)_\mu$ and $a\mu = \{s \in [0,T): \beta(\bar{\mu}) = \varepsilon(s,\mu)\}$. Then for each $\mu > 0$ small the two sets $A\mu$, $a\mu$ are either both empy or both non empty and in a one to one correspondence.

We explore now the stability properties of the nonzero periodic solutions of $(3.2)\mu$ by analyzing the stability behaviour of the equilibrium points of $\pi = \pi_{\mu}^{t_\circ}$ relative to any fixed $t_\circ$. Clearly if γ_{μ} is a periodic trajectory and p is the corresponding fixed point of π, then the stability behaviours of γ_{μ} under $(3.2)_{\mu}$ and of p under π are the same.

Let $\tilde{\pi}$ be the restriction of π to $M\mu(t_\circ)$. In connection with (3.3) consider the function $F(t_\circ,\theta_\circ,r_\circ,\mu) = \theta(t_\circ+ T,t_\circ,\theta_\circ,r_\circ,\mu)- \theta_\circ - 2\pi$ and its restriction $\tilde{F}(t_\circ,\theta_\circ,\mu) = F(t_\circ,\theta_\circ,g_{\mu}(\theta_\circ,t_\circ),\mu)$ $t_\circ$ $M\mu(t_\circ)$. The fixed points of $\tilde{\pi}$ correspond to the zeros of $\tilde{F}$. We suppose that the set Ω of these points is finite, $\Omega = \{p_1,\ldots,p_N\}$. It is not difficult to prove that any arc of M $(t_\circ)$ whose endpoints are in Ω is invariant under $\tilde{\pi}$.

This implies some relationships between the number N and the stability behaviour under $\tilde{\pi}$ of the points in Ω. For instance, if $N = 1$, p_1 is semistable and is also a nonuniform attractor. If $N > 1$ each p_j which is an attractor is also a uniform attractor, and each p_j which is semistable cannot be an attractor. The negative attractivity of a fixed point may occurr only when $N > 1$ and is always uniform. Uniform attractors and negative attractors are in the same number. Then if N is odd, at least one fixed point is semistable.

The attractivity properties of $p_1,\ldots,p_N$ under $\tilde{\pi}$ can be expressed in terms of $\tilde{F}$. Precisely if $p = (\theta_\circ,g_{\mu}(\theta_\circ,t_\circ)) \in \Omega$, then: (i) p is semistable if and only if $\tilde{F}$ has a local proper extremum at $\theta_\circ$; (ii) p is a uniform attractor (resp. a negative attractor) if and only if $\tilde{F}$ is strictly decreasing (resp. increasing) in a neighbourhood of $\theta_\circ$. More generally $\tilde{F}$ contains all the informations concerning the attractivity properties of $p_1,\ldots,p_N$ with respect to the unrestricted dynamical system π. Indeed the following theorem holds.

Theorem 4.2. We have: (i) each point $q \in H(t_\circ)\setminus\{0\}$ is attracted under $\pi = \pi_\mu^{t_\circ}$ to one of the points p_j; (ii) each p_j which is a uniform attractor under $\tilde{\pi}$, is a uniform attractor (i.e. asymptotically stable) under π.

Theorem 4.2. may be paraphrased as follows. Let $\gamma_1,\ldots,\gamma_N$ be the periodic solutions of $(3.2)_\mu$ which correspond to $p_1,\ldots,p_N$. Then: (i) each solution of $(3.2)_\mu$ starting in $H\setminus M_\circ$ approaches one of the γ_j; (ii) any s_j which is asymptotically stable on $M\mu$ is unconditionally asymptotically stable.

We conclude by the following characterization of the stability properties of periodic solutions in terms of the function $\varepsilon(s,\mu)$ of Theorem 4.1.

Theorem 4.3. Assume h = 3. Define $A\mu$ and $a\mu$ as in Theorem 4.1 and suppose that $A\mu$ is finite. If $\gamma \in A\mu$ and s is the corresponding number in $a\mu$, we have: (i) γ is stable (and then asymptotically stable) if and only if $\varepsilon(\cdot,\mu)$ is strictly decreasing in a neighbourhood of s; (ii) when γ is unstable then the intersection of γ with any section $M\mu(t_\circ)$ is a repellor or a semistable point of $\pi_\mu^{t_\circ}$ and is a repellor if and only if $\varepsilon(\cdot,\mu)$ is strictly increasing in a neighbourhood of s.

References

[1] Bernfeld, S.R., Salvadori, L. and Visentin, F., Hopf bifurcation and related stability problems for periodic differential systems, J. Math. Anal. and Appl. 116, (1986), pp. 427-438

[2] Chow, S.N. and Hale, J.K., Methods of bifurcation theory, Springer Verlag, New York/Berlin, 1982

[3] Kelley, A., The stable, center-stable, center-unstable, and unstable manifolds, J. Diff. Eq.

[4] Lanford, O.E., Bifurcation of periodic solution into invariant tori: The work of Ruelle and Takens, Lecture Notes, 322, Springer Verlag, Berlin, 1972

[5] Salvadori, L., Bifurcation and stability problems for periodic differential systems, Nonlinear oscillations for nonconservative systems, Editor A. Ambrosetti, Pitagora, Bologna, 1985, pp. 93-104.

FORCED OSCILLATIONS OF HAMILTONIAN SYSTEMS

A. Salvatore
Dipartimento di Matematica - Università di Bari
Via G. Fortunato, Campus, 70125 Bari
ITALY

0. INTRODUCTION

We shall look for T-periodic solutions of the non-autonomous Hamiltonian system

$$(0.1) \qquad \begin{aligned} \dot{p} &= -H_q(t,p,q) \\ \dot{q} &= H_p(t,p,q) \end{aligned}$$

where $H \in C^1(\mathbb{R}^{2n+1}, \mathbb{R})$, $p,q \in \mathbb{R}^n$, denotes $\frac{d}{dt}$, $H_p = \frac{\partial H}{\partial p}$, $H_q = \frac{\partial H}{\partial q}$ and $H(t,z)$ is T-periodic in t. Many authors have studied the problem of the existence of free and forced oscillations of (0.1) under different assumptions of the growth of H at infinity (cf. e.g. the review article [18] and its references).

More precisely, they assume that H is superquadratic in z (resp. subquadratic or asymptotically quadratic) i.e. $H(t,z)/|z|^2 \to +\infty$ as $|z| \to +\infty$ (resp. $H(t,z)/|z|^2 \to 0$ or $H(t,z) = \frac{1}{2}(b_\infty(t)z,z)+G(t,z)$ where $b_\infty(t)$ is a symmetric matrix for any $t \in [0,T]$ and $G(t,z)/|z|^2 \to 0$ as $|z| \to +\infty$.

However the above assumptions on the growth on H do not cover an important class of Hamiltonian functions which occurs in the study of mechanical systems, namely the Hamiltonian function of the type:

$$(0.2) \qquad H(t,p,q) = \tfrac{1}{2} \sum_{i,j=1}^{n} A_{ij}(t,q)p_i p_i + V(t,q)$$

where $\{A_{ij}(t,q)\}$ is a positive definite matrix and $A_{ij}(t,q)$ and $V(t,q)$ are C^1 real valued functions.

Hamilton functions of the form (0.2) has been considered in [4], [5], [11], [19] in the case in which $V(t,q) \to +\infty$ as $|q| \to +\infty$.

Now we shall study the existence of forced oscillations of (0.1), where H is as in (0.2), in two very interesting situations:

i) the potential $V(t,q)$ is bounded

ii) the potential $V(t,q)$ is singular.

To study these cases it will be convenient to consider the Lagrangian formulation of (0.1).

Namely, since H in (0.2) is convex in the variable p, it is well known that the T-periodic solutions of (0.1) correspond, in a suitable sense, to the T-periodic solutions of the Lagrangian system of differential equations

$$(0.3) \qquad \frac{d}{dt}\frac{\partial L}{\partial \xi}(t,q,\dot{q}) - \frac{\partial L}{\partial q}(t,q,\dot{q}) = 0$$

where

$$(0.4) \qquad L(t,q,\xi) = \tfrac{1}{2} \sum_{i,j=1}^{n} a_{ij}(t,q)\xi_i \xi_j - V(t,q)$$

is the Legendre transform of $H(t,p,q)$ respect to p, and $\{a_{ij}(t,q)\}$ is the inverse matrix of $\{A_{ij}(t,q)\}$ (cf. [16]).

In sections 1 and 2 we shall give a partial answer to the problem of the existence of forced solutions of (0.3) in the cases i) and ii).

1. THE CASE OF THE BOUNDED POTENTIAL

Consider the forced Lagrangian system

$$\frac{d}{dt}\frac{\partial L}{\partial \xi}(q,\dot{q}) - \frac{\partial L}{\partial q}(q,\dot{q}) = g(t) \tag{1.1}$$

where g(t) is a T-periodic "forcing" term. Then

$$L(t,q,\xi) = \frac{1}{2}\sum_{i,j=1}^{n} a_{ij}(t,q)\xi_i\xi_j - V(t,q) - g(t),q) \tag{1.2}$$

where (,) denotes the inner product in $\mathbb{R}^n$.

In [8] we have stated the following theorem:

Theorem 1.3. Assume g(t) is an integrable T-periodic function with zero mean value. Moreover

B_1) there exists $\lambda > 0$ s.t. $\lambda|\xi|^2 \leqq \sum_{i,j=1}^{n} a_{ij}(q)\xi_i\xi_j$ for any $q,\xi \in \mathbb{R}^n$

B_2) V(q) is bounded

B_3) there exists $\tau > 0$ s.t. $a_{ij}(q)$ and V(q) are τ-periodic.

Then (1.1) has at least two T-periodic solutions which do not differ by a multiple of τ.

Remark 1.4. This theorem can be applied to the study of the forced oscillations of the double pendulum. It generalizes some results proved by Mawhin and Willem for the forced pendulum equation (cf. [17] and [20]).

Remark 1.5. If $a_{ij}(q)$ = const and V(q) bounded, we have studied this problem in the resonant case (cf. [7]).

Now we shall give a sketch of the proof. Let be $H^1 = H^1([0,T],\mathbb{R}^n)$ the Sobolev space obtained by the closure of the C^∞ T-periodic functions q(t) with respect to the norm

$$\| q \| = (\int_0^T |q(t)|^2 dt + \int_0^T |\dot{q}(t)|^2 dt)^{\frac{1}{2}} .$$

It is easy to see that the T-periodic solutions of (1.1) are the critical points of the functional

$$(1.6) \qquad f(q) = \tfrac{1}{2} \int_0^T \sum_{i,j=1}^{n} a_{ij}(q)\dot{q}_i\dot{q}_j dt - \int_0^T V(q)dt - \int_0^T (g,q)dt$$

Obviously f is bounded from below. In order to find the critical points of f using variational methods, we need to verify the Palais-Smale condition ((PS) condition).

Unfortunately, since V is bounded, the (PS) condition in general does not hold: however the periodicity assumption implies a "suitable" (PS) condition.

Let be $\tilde{H}^1 = \{q \in H^1 | \int_0^T q = 0\}$. Then

for any $q \in H^1$ $q = q^\circ + \tilde{q}$, where $q^\circ \in \mathbb{R}^n$ and $\tilde{q} \in \tilde{H}^1$.

It is possible to prove that f satisfies the (PS)* condition, i. e. "given $c \in \mathbb{R}$, every sequence $\{q_k\}$ for which

$$f(q_k) \to c \ , \quad f'(q_k) \to 0$$

and

$$q_k = q_k^\circ + \tilde{q}_k, \ q_k^\circ \in [0,\tau]^n \text{ for any } k \in N ,$$

possesses a converging subsequences".

Since f is τ-periodic, the (PS)* condition implies a classical deformation lemma, then by standard arguments it follows that $\beta = \inf_{H^1} f$ is a minimum of f. Therefore system (1.1) has a solution p which minimizes f on H^1.

In order to find a different solution, we shall apply the Ambrosetti-Rabinowitz mountain pass lemma (cf. [2]).

Assume that p is a strict local minimum of f: then there exist $R, \rho > 0$, $\rho > \beta = f(p)$ such that

$$f(q) \geqq \rho \quad \text{for any } q \in H^1, \ \| q-p \| < R .$$

The τ-periodicity of f implies that $R < \sqrt{T\tau}$. Moreover

$$f(p + \tau) = f(p) < \rho \qquad \text{with } \| p + \tau - p \| = \sqrt{T\tau}$$

By mountain pass lemma, there exists a critical value $\beta_1 \geqq \rho > \beta$; obviously the corresponding critical point is a solution which cannot differ from p by a multiple of τ.

<u>Remark 1.7</u>. Instead of (PS)* condition, we could use the $(PS)_c$ condition of Brézis, Coron and Nirenberg and a slightly different version of the mountain pass lemma (cf. [6], [17]).

2. THE CASE OF THE SINGULAR POTENTIAL

Let us consider a Lagrangian system in presence of singularities, i.e. with a potential V(t,q) defined on $\mathbb{R} \times \Omega$, Ω open subset of $\mathbb{R}^n$, and $V(t,q) \to -\infty$ as $q \to \partial\Omega$. This problem has been studied by W.B. Gordon (we refer to [12], [13] also for the physical motivation of the problem). Moreover we refer to [3], [14] for the case $V(t,q) \to +\infty$ as $q \to \partial\Omega$ and to [1] for the convex case (cf. also the paper of Ambrosetti in this volume).

Following [9], in this section we suppose that $L(t,q,\xi)$ has the form (0.4) and

$$(S_1)\quad \begin{cases} \{a_{ij}(t,q)\xi_i\xi_j\} \text{ is a positive definite matrix, i.e. there exists} \\ \text{a function } \lambda : \mathbb{R}^n \to]0, +\infty[\text{ such that:} \\ \text{(i)} \quad \sum_{i,j=1}^{n} a_{ij}(t,q)\xi_i\xi_j \geqq \lambda(q)|\xi|^2 \quad \text{for any } t \in \mathbb{R},\ q,\xi \in \mathbb{R}^n \\ \text{(ii)} \quad \text{there exist real constants } c_1 > 0 \text{ and } \upsilon \in [0,1] \text{ such that} \\ \qquad \lambda(q) \geqq c_1(|q|^{\upsilon} + 1)^{-1} \text{ for any } q \in \mathbb{R}^n; \end{cases}$$

there exist a function $U \in C^1(\Omega, \mathbb{R})$, a neighbourhood $\mathcal{N}$ of $\mathbb{R}^n \setminus \Omega$ and a constant $c_2 \geqq 0$ such that

(S_2) (i) $\lim_{q \to \partial\Omega} U(q) = -\infty$

(ii) $-V(t,q) \geqq |U'(q)|^2 - c_2$ for any t $\mathbb{R}$ and $q \in \mathcal{N} \cap \Omega$

In [9] we have proved these theorems, which have been announced in rather different version in [10] (cf. also [15] for some generalizations).

Theorem 2.1. Let be $N = 2$ and $\Omega = R^2 \setminus \{0\}$. Assume (S_1), (S_2) and

(S_3) there exist real constants c_3, $c_4 > 0$ and $\mu \in [0, 2 - \nu[$ s.t. $V(t,q) \leqq c_3 |q|^{\mu} + c_4$ for any $t \in R$, $q \in \Omega$.

Then there exists at least one T-periodic solution of (0.3)

Theorem 2.2. Let $N \geqq 2$ and let Ω be symmetric with respect to the origin. Assume (S_1), (S_2), (S_3) and

(S_4) $a_{ij}(t,q)$ and $V(t,q)$ are T/2-periodic in t and even in q.

Then there exists at least a pair $(q, -q)$ of T-periodic solutions of (0.3).

In order to prove these theorems, let us consider the functional of the action (1.6) which is defined on the open subset

$\Lambda^1\Omega = \{q \in H^1 | q(t) \in \Omega$, for $t \in [0,T]\}$ of H^1.

It is easy to verify that f is weakly lower semicontinuous and bounded from below on a suitable subset of $\Lambda^1\Omega$. More precisely, if $\Omega = \mathbb{R}^2 \setminus \{0\}$, we restrict f to the class of the functions of $\Lambda^1\Omega$ which are not homotopic to a constant on Ω, if $N \geqq 2$ and Ω is symmetric, we consider $f|_E$,

$$E = \{q \in H^1 | q(t+T/2) = -q(t)\} .$$

By the assumptions (S_2) and (S_4) the minimum of f on these classes is attained at a point q_0 which belongs to $\Lambda^1\Omega$ and is a solution of (0.3).

Remark 2.3. Assumption (S_2) has been introduced by W.B. Gordon to study the autonomous second order Hamiltonian system $\ddot{q} = -V(q)$ with V(q) bounded from above. For example, if $V(tq) = -\frac{1}{|q|^2} + (g(t),q)$, then (S_2) holds with

$$U(q) = \log |q|, \quad \mathcal{N} = \{q \in \mathbb{R}^2 | \ |q| < 1\} \quad \text{and} \quad c_2 = |f|_\infty .$$

3. REFERENCES

[1] Ambrosetti, A., Coti Zelati, V.: "Solutions with minimal period for Hamiltonian systems in a potential well", preprint

[2] Ambrosetti, A., Rabinowitz, P.H.: "Dual variational methods in critical points theory and applications", J. Funct. Analysis, 14 (1973), 349-381

[3] Benci, V.: "Normal modes of Lagrangian systems constrained in a potential well", Ann. Inst. H. Poincaré, 1, 5 (1984), 379-400

[4] Benci, V., Capozzi, A., Fortunato, D.: "Periodic solutions for a class of Hamiltonian systems", Springer Verlag Lectures Notes in Mathematics, 964, (1982), 86-94

[5] Benci, V., Capozzi, A., Fortunato, D.: "Periodic solutions of Hamiltonian systems with superquadratic potential", to appear on Ann. Mat. Pura e Appl.

[6] Brézis, N., Coron, J.M., Nirenberg, L.: "Free vibrations for a nonlinear wave equation and a theorem of P. Rabinowitz", Comm. Pure Appl. Math., 33 (1980), 667-689

[7] Capozzi, A., Fortunato, D., Salvatore, A.: "Periodic solutions of dynamical systems", Meccanica, 20 (1985), 281-284

[8] Capozzi, A., Fortunato, D., Salvatore, A.: "Periodic solutions of Lagrangian systems with bounded potential", to appear on J. Math. Anal. Appl.

[9] Capozzi, A., Greco, C., Salvatore, A.: "Lagrangian systems in presence of singularities", preprint

[10] Capozzi, A., Salvatore, A.: "Periodic solutions of Lagrangian systems: the case of the singular potential", Proceedings of the NATO ASI, (1986), 207-216

[11] Giannoni, F.: "Soluzioni periodiche di sistemi Hamiltoniani in presenza di vincoli", Pubblic. Dip. Mat. Univ. Pisa, 11, (1983)

[12] Gordon, W.B.: "Conservative dynamical systems involving strong forces", Trans. Amer. Math. Soc., 204, (1975), 113-135

[13] Gordon, W.B.: "A minimizing property of Keplerian orbits", Amer. J. Math., 99 (5), (1977), 961-971

[14] Greco, C.: "Periodic solutions of second order Hamiltonian systems in an unbounded potential well", to appear on Proc. R. Soc. Edinb.

[15] Greco, C.: Periodic solution of a class of singular Hamiltonian systems", preprint

[16] Landau, L.D., Lifsits, E.M.: "Meccanica", Editori Riuniti, Edizioni Mir, (1976)

[17• Mawhin, J.,Willem, M.: Multiple solutions of the periodic boundary value problem for some forced pendulum-type equations, J. Differential Equations, 52, 2, (1984) 264-287

[18] Rabinowitz, P.H.: "Periodic solutions of Hamiltonian systems: a survey", SIAM J. Math. Anal. 13, (1982)

[19] Salvatore, A.: "Periodic solutions of Hamiltonian systems with a subquadratic potential, B.U.M.I.", (C), 1, (1984), 393-406

[20] Willem, W.: "Oscillation forcées de systêmes Hamiltonian", Publications Sémin. Analyse non linéaire, Univ. Bésancon (1981).